Hierarchical Structures

Participants

Reading from left to right:

BACK ROW
E. Harrison, A. Wachman, A. Wilson, C. Smith, R. Saunders, G. Kocher, M. Mesarovic, R. Rosen, P. Shlichta, S. Enzer, C. Leake, I. Haissman, P. Pearce, R. Williams, H. Gutman, L. Larmore.

THIRD ROW
C. Perry, R. Lucky, B. Wahl, V. Gradecak, J. Zimmerman, H. Wolbers, W. Parkyn, A. Gralnik, C. Alexander, F. Tonge, J. Gauger.

SECOND ROW
D. Wilson, D. Macko, M. Kaufman, R. Jones, A. Mood, L. Whyte, M. Juncosa, J. Dieges, M. Bunge, R. Specht, V. Azgapetian, K. Justice, M. Lodato.

FRONT ROW
P. Gerrard, H. Pattee, R. Gerard, M. Grene, M. Maruyama, T. Page, A. Goodman.

PROCEEDINGS
of the
SYMPOSIUM
held
NOVEMBER 18-19, 1968
at
DOUGLAS
ADVANCED RESEARCH
LABORATORIES
Huntington Beach
California

Hierarchical Structures

Edited by

Lancelot Law Whyte
Albert G. Wilson
Donna Wilson

American Elsevier
Publishing Company, Inc.
NEW YORK · 1969

AMERICAN ELSEVIER
PUBLISHING COMPANY, INC.
52 Vanderbilt Avenue, New York, N.Y. 10007

ELSEVIER PUBLISHING COMPANY, LTD.
Barking, Essex, England

ELSEVIER PUBLISHING COMPANY
335 Jan Van Galenstraat, P. O. Box 211
Amsterdam, The Netherlands

Standard Book Number 444-00069-0

Library of Congress Card Number 76-99092

Manufactured in the United States of America

Preface

This book is based on the interdisciplinary symposium, "Hierarchical Structure in Nature and Artifact," held in November 1968 at Huntington Beach, California. The symposium was convened by Lancelot Law Whyte and Albert Wilson under the sponsorship of the Douglas Advanced Research Laboratories and the University of California, Irvine to bring together scientists, engineers, designers, and others interested in the function of hierarchical structures in nature, concept and design. Through placing in juxtaposition specific hierarchical systems from the inorganic, organic, conceptual, and artifact worlds, it was hoped to gain insight into the problems of levels, parts and wholes, and the origin of the various species of hierarchical structures.

For purposes of the symposium, the terms "hierarchical structure" and "hierarchy" were taken generally to mean *a set of ordered levels.* Whereas a more orthodox definition of "hierarchy" requires a *governing-governed* relation between levels, this attribute was intended only when specified. It was felt that this symposium, the first scientific gathering on *hierarchy,* should explore rather than define. Consequently, it was decided to postpone sharpening of terminology until the full variety of meanings given to the term "hierarchy" could be assimilated. For this reason, a standard terminology is not used throughout this book. However, this causes little confusion, since most of the authors are careful to amplify the meaning of the terms they introduce.

Beyond the questions of definition and classification, several basic problems concerning hierarchical structures were raised: Do some or all of the hierarchies we discern in nature possess objective reality or are they subjective patterns derivative from the human mode of perception and conception; if levels are structural realities, can the origin of inorganic hierarchies be explained in terms of known physical laws without improbable *ad hoc* initial conditions; can a reductionist explanation be found for the levels of biological organization; do the similarities between the various species of hierarchies and level structures imply a structural commonality that is meaningful on some level of abstraction; if so, can the existence of such structures be derived from some fundamental meta-principle – informational, combinatorial, topological, or whatever. These and other relevant questions were approached during the symposium along a path leading from the specific to the general. While few answers were forthcoming, the new differentiations and syntheses developed by the participants gave the general feeling that the proceedings produced much of value to the embryo subject of hierarchical structure.

The material generated for and by this symposium on hiearchies appears here in the form of the papers invited to be read at the symposium and notes based on the ensuing discussions. Instead of publishing the verbatim discussion following the presentation of each paper, the editors invited those making substantial contributions to the discussion to prepare brief formal notes. These have been included at the end of each topical part. In addition to the papers and notes, a selected annotated bibliography coverning a sizeable portion of the existing literature on hierarchies has been included.

The editors hope that this volume will provide a useful overview for those who have an interest in the problems of levels, hierarchies, parts and wholes, reductionism, holisim, and general systems whatever the area of application. Finally, we also hope that the synoptic material covered in this book will further erode disciplinary overspecialization and lead to the creation of a new fraternity of communication.

■

Acknowledgements

The editors want to offer their sincere thanks to Dr. Lewis Larmore and the McDonnell Douglas Corporation for support of the symposium and this publication, and to Dean Ralph Gerard for the support of the University of California, Irvine. Cordial thanks are also given to the session chairmen, Victor Azgepetian, Ralph Gerard, Donald Menzel, and Paul Shlichta and to Mario Bunge, chairman of the post session workshop, for their personal contribution and for promoting lively and penetrating exchanges. We extend special thanks to Joe Gauger, Jeanne Gray, and Thornton Page for their assistance in logistics and coordination. We acknowledge our indebtedness to the Charles Eames Staff for their film, *The Powers of Ten*. We are pleased to acknowledge the splendid cooperation and support of Jim Eastman and his enthusiastic staff of McDonnell Douglas Astronautics Company – Western Division, with special gratitude to Robert Fisher and Robin Simpson in helping prepare the camera ready copy from which this volume was produced. Finally we want to thank the authors and all other participants who by their contributions have opened doors on the understanding of hierarchical structures.

■

Contents

PART III Organic Hierarchical Structures

PART IV Hierarchy in Artifact

Part I

Hierarchy in Concept

As humans, we belong to that component of nature given to organizing and structuring. We not only physically organize ourselves and our environment, but we also organize our perceptions of the physical world into abstract structures. When we project these abstractions back onto the physical world, their usefulness leads us to surmise that they reflect to some degree a structure possessing independent existence.

The human method of conceptualization discriminates entities, relations, processes, and levels as the ingredients of structure. The scientific study of structures and systems – natural, artificial, or abstract – has primarily been concerned with entities, relations, and processes largely ignoring the roles of levels and hierarchies because of their complexity. However, Lancelot Law Whyte in documenting the history of thought concerning hierarchical structure from Plato and Aristotle to the twentieth century establishes in the first paper of Part I the thesis that the study of hierarchies has now come of age. As we engage in the study and creation of structures and systems of larger complexity, the essential role of levels and hierarchies in complex situations is increasingly realized as is evidenced by the current expansion in the literature of many disciplines which treats this subject.

Mario Bunge in the second paper suggests some useful working definitions for the concepts of *hierarchy*, *level structure* and *level*. Bunge's basic definition is that of a level structure which is taken as a family of sets, having a relation between the sets that represents emergence or a novelty generating process. The emergence relation that holds *between* the sets does not hold *within* the set whose elements are taken to be qualitatively homogeneous systems. Bunge defines a *level* as a set having these properties and belonging to a level structure. If, instead of the emergence relation between sets or levels, there is an anti-symmetric dominance relation, the level structure is a *hierarchy*. Bunge develops the ontological and epistemological aspects of structures with these properties.

M. D. Mesarovic and D. Macko consider three concepts of hierarchy: (i) Hierarchies of description whose levels (called strata) are of description or abstraction; (ii) Multi-layer decision systems whose levels (called layers) are sequential events in a decision making process; and (iii) Multi-level multi-goal systems whose levels (called levels) are those of an organizational hierarchy. In the first concept there is autonomy of language and principle on each strata, but an asymmetrical interdependence of function between different strata. In the second concept each layer specifies constraints affecting the operation of subsequent layers. In the third concept interacting subsystems are structured to develop capability for tasks beyond the capacity of individual units.

Amplification of the discussion of the concepts of hierarchy is contained in four brief notes. Lancelot Law Whyte raises five primary questions pertaining to the properties and origin of structural hierarchies. Robert Rosen stresses that the interaction between the functional levels in a biological hierarchical system are reciprocal relations and not unidirectional, although the possibility of a pair of "bossing" relations operating in opposite directions exists. Albert Wilson describes the role of topological and temporal closure in defining levels in inorganic hierarchies. Marjorie Grene, in searching for a unifying concept in the different usages of hierarchy, suggests that levels are always governed by some form of ordering relation.

■

Structural Hierarchies: A Challenging Class of Physical and Biological Problems

Lancelot Law Whyte*

An astronomer at home with galaxies and a natural philosopher interested in atoms initiated this symposium. This collaboration is not arbitrary for it reflects the concern of some physical scientists to link macrocosm and microcosm, by finding parameters and equations linking the ultra small and the unimaginably huge. This vaulting ambition, seeking to pass from the minute to the immense at one leap, may well prove fertile. But it must be supplemented by co-operation between specialists so that exact science can advance level by level through the ranges of this graduated universe, tracing the relations of all the levels in the great hierarchies of structure from the smallest units to the largest. This step by step procedure may prove the most reliable path towards understanding of this complex structured universe, with its opposed tendencies towards disorder and towards order.

Why is 1968 appropriate for this symposium? Actually *two* symposia on hierarchy were held in 1968: Koestler's in Austria in August, and ours in California in November! This widespread sense of timeliness may be due to the fact that it has recently been realized that workers in several fields are now occupied in studying the character of complex units and the quantitative aspects of hierarchical structures, for example in cosmology, in protein molecule conformation, in speculative theories of nuclear structure and elementary particle classification, and in the formal structures of codes, computer languages, and decision making systems. Finally, as though to mark this as the decade of hierarchy, since 1962 a cluster of significant essays or books on hierarchy have appeared,e.g., those of Bernal (1958, 1962), Simon (1962), Ando (1963), Layzer (1964), Harris (1965), Smith (1965), Wilson (1965, 1967), Grene (1967), Koestler (1967), myself (1949, 1965, 1969) – and surely there are others, perhaps in Soviet Russia. This year is appropriate because many scientists realize that the moment has arrived

**93 Redington Road, London N. W. 3, England.*

when exchanges between students of structural hierarchies in different realms, not only in physics, biophysics, cosmology, and artifacts, but also in the psychological and social sciences, will be of mutual benefit. Some sixty years of the study of structure have sufficed to make it obvious, at least to many younger minds, that it is time for closer attention both to the precise objective character of ordered units and to the theoretical significance of hierarchical ordering of various kinds.

The history of science is part of science. An idea is only fully understood when we know enough about its history to see it as a stage in the continuing advance of science, from yesterday through today to conjectures about its significance tomorrow. It is therefore appropriate that out of the fourteen papers in this volume, three – those of myself, Leake, and Gerard – are historically oriented.

But before we plunge into the river of history, with its stagnant phases followed by cataracts of discovery, two prior issues arise. Is not hierarchical classification practically universal? If so, what precisely do we at this symposium mean by a spatial, or better, a *structural hierarchy*?

The immense scope of *hierarchical classification* is clear. It is the most powerful method of classification used by the human brain-mind in ordering experience, observations, entities, and information. Though not yet definitely established as such by neuro-physiology and psychology, hierarchical classification probably represents the prime mode of co-ordination or organization (i) of cortical processes, (ii) of their mental correlates, and (iii) of the expression of these in symbolisms and languages. As a reminder of its great scope I cite the hierarchical classification of:

- Numbers, scales, times, positions, crystal forms, symmetry elements, and groups.

- Symbolisms, sentences, and languages of all kinds.

- Logical types, concepts, principles, information, quantities, and abstractions of many other sorts.

The use of hierarchical ordering must be as old as human thought, conscious and unconscious, and the use of hierarchical classification in artifacts such as codes, computers, and the like, is the theme of one of the four sessions of this symposium. The other three are concerned with hierarchical structures in nature, that is in physics, in biophysics and general biology, and in cosmology.

But should we not define at the outset what we are talking about? That would be all right if we knew! As it is, we have still to identify the precise structure of natural hierarchies, what properties characterize them in general, and what differences arise in different realms and different circumstances.

In this dilemma it is sometimes considered that two alternatives are open. One is to start with verbal definitions of logically sharp *ideas*, even if we suspect that they are likely to prove inappropriate. The other is to get on with the study of the *facts*, though we may realize that what we see as "facts" may be distorted by a bias in our way of thinking. But these are not valid alternatives, for we use both ideas and facts all the time. Even if it is uncomfortable for philosophers of science hasty for clarity, what we willy nilly have to do in pursuing intellectural clarity and unification, is continually to rely on a trained intuitive and imaginative *judgment*. Our imaginative judgment has to be used at every step to see improved ideas latent in the facts, and to identify the new facts which may follow from improved ideas that are not yet logically clear. In this symposium we try to give both aspects their due: the ideal of logical clarity and the stubborn kernel of fact from which subjective prejudices have been removed, as far as possible. Our exchanges should assist that difficult task.

Let me attempt a first preliminary formulation. We are agreed, I will assume, that the entities of science do not all enjoy equal empirical and theoretical status, but are sometimes arranged in sequences of discrete separable identifiable levels, from "higher" to "lower," e.g., in structural hierarchies from larger to smaller units. I will risk a further step, and assume, as some biologists such as Paul Weiss (1967) have, that in some manner (probably not yet fully understood at any level and certainly not at many levels) the existence of a system or unit at one level in some cases imposes constraints on the degrees of freedom of the parts. One can therefore posit, as a working hypothesis, that (at some levels, in organic hierarchies, under certain unknown conditions):

> A structural level, or unit, in a structural hierarchy can be usefully defined (in a special or restricted sense) as "a three-dimensional system of parts, (i.e., structures or processes) involving (within certain thresholds) characteristic constraints imposed on the degrees of freedom of its first-order parts, so that the properties of the unit are not the simple linear summation of the properties which the same parts display when isolated (i.e., when they are not subject to those constraints)." The new properties of the higher level, e.g., in those of a newly formed unit, may arise in at least two manners: (i) when, in a more or less random system of interacting units, a global threshold is passed, e.g., when a global potential energy function becomes greater than the total of the two-term interactions, resulting in "synthesis" of the new form, "clustering," etc.; or (ii) when, within a partly ordered unit or a system under global constraints, local thresholds are passed, so that "fragmentation," "dispersal of a unit," etc., occurs.

This is an appropriate principle, supported by some evidence. But we are still too ignorant about the levels in different kinds of hierarchies to reach firm conclusions. This suggestion merely indicates a possibly fertile line of quantitative analysis.

With that preliminary clarification of what we are discussing, I return to the history of the idea of hierarchy, and of structural hierarchy in particular. Lest we interpret the historical advance too naively, for example by isolating physics, I start with two quotes, both from logicians. Petrus Ramus (1515-1572) a

French humanist,[1] wrote: "Everything is composed of little units, and the mind groups these spatially into clusters, or breaks such clusters down into their units." Note the parallel with the cosmologists' objective processes of "clustering" and "fragmentation." Even more enlightening as it is in advance of much current writing on hierarchy, is this from Albert Ando (1963):

> "The possibility of identifying 'causal' relations is intimately related to the possibility of classifying variables into a hierarchy of sets, levels I, II, III, and so on. Variables belonging to higher-numbered sets are influenced by those in the lower-numbered sets, but the former do not influence the latter. . . . when such a stratification exists, then we may say that the variables in the lower-numbered sets are the 'causes' of the variables in the higher-numbered sets. This type of hierarchical structure also provides the justification for ignoring the variables in higher-numbered groups when the object of an investigation is restricted to the behavior of variables in lower-numbered sets. *Laboratory experimentation is a technique for artifically creating a particularly simple form of hierarchy among variables."* (my italics)

It is not surprising that this penetrating observation should come from a Japanese-born economist, for it is easier to recognize the determining or limiting conditions of a theory, or of a scientific method, from well outside it. Even proud physics can gain from the study of the past and from the observations of an economist! There is a continuity and unity in knowledge.

Lest this be thought an arbitrary chance, let me cite another example: William Stern,[2] a German sociologist, psychologist, and philosopher, wrote *in 1923,* "There is a ranked system ("Stufen-system") of equations of very different orders of scope ("Weitenordnung") from the energy-equivalences, which exist between all physical functions, down to the uniformity which is displayed in the particular pigmentation of the hair of

1. See *Ramus, Method and the Decay of Dialogue* by W. J. Ong 1958, Cambridge: Harvard University Press.

2. *Person und Sache. Volume I: Ableitung und Grundlehre* Leipzig: J. A. Barth 1906, p. 356. See also Vol II: *Die menschliche Persönlichkeit* 1918, p. 8. (In my copy of Vol II, I find a note on "hierarchy" made c. 1925. I am indebted to Herbert Gutman for reminding me of Stern's interest in hierarchy.)

an individual because this hair colour is an element in the permanent characteristics of the individual."

The term "hierarchy" is said to derive from Pseudo-Dionysius.[3] The idea runs from Plato,[4] Aristotle,[5] Pseudo-Dionysius, with his angelic and priestly hierarchy, through mediaeval philosophers to the 15th Century thinkers of the Florentine Academy[6] who began to loosen up the largely static classical hierarchy, and many German thinkers,[7] some with a *Schichten-lehre,* to Shapley (1930), for example, in our own century. I refer you to Lovejoy's *Great Chain of Being (1936)* and to a brief historical survey which I am publishing elsewhere (Whyte 1969). Towards the end of the 18th Century the growing influence of Galileo, Kepler, Descartes, Newton, and their followers, blew away the scholastic hierarchies of angels, priests, and substantial forms, and thus left the field clear for heirarchy to stage a comeback in our century, as a type of dynamic and sometimes unstable spatial ordering which is recognized to pervade the universe (though still neglected by some sciences, for neither the term nor the explicit idea is to be found in most physics texts). Since around 1950, Plato's static hierarchy has come back, often as an unstable hierarchy of *processes.*

3. See article by E. F. Osborn in *Encyclopedia of Philosophy*. ed. Edwards. New York: Macmillan. 1967.

4. Plato's many Dialogues treating hierarchical conceptions (ranking of Idea, Soul as series of levels, etc.) cannot be detailed here. See N. Hartman, "Die Anfänge des Schichtengedankens in der Alten Philosophie." Abh. d. Pr. Ak. d. w. Berlin 1943. Phil. Hist. No. 3, also note 7.

5. Many of Aristotle's writings express his conception of a single great, yet continuous hierarchy. See particularly "On Hierarchical Structure of Nature," *Hist. Anim VIII*. 1. pp. 588-9 in Berlin Akad. Edition. Bekker, 1831. The hierarchical ideas of Plotinus, Thomas Aquinas and others cannot be included here.

6. See P. O. Kristella on M. Ficino (1433-99) in *Encyclopedia of Philosophy*. ed. Edwards. New York: Macmillan. 1967. Casanus and Pico are also relevant.

7. For German articles and references on *Schichtenlehere* (doctrine of layers or levels) from Plato and Aristotle to N. Hartman, see *Studium Generale* (9 Jahrgang, Heft 6, July (1956) Springer. These essays and Hartman's writings (1912 onwards) are indispensable material for a history of the concept of hierarchy, though set in particular philosophical contents different from that of contemporary science.

But is it a come-back? Did all mathematical physicists through the centuries from Galileo and Kepler to Einstein and Bohr entirely neglect this powerful idea? The answer is as one would expect: exceptionally imaginative minds recognized its power and amused themselves by trying to apply it. I take four examples: Newton (1705), Lambert (1761), Fournier d'Albe (1907), and Charlier (1908) made quantitative conjectures (rather close to the mark though the values of their parameters were not specified) about a hierarchical structuring of the physical universe. The idea of structural hierarchy was too interesting and fertile not to play with when *heterogeneity* had to be represented, and science is kept alive by intelligent conjectures of this kind. Indeed few geometrical ideas are more exciting than that of ordered *hierarchical arrangement*, beside which the uniform close-packing in linear arrays is a relatively simple idea, relevant only to homogeneous systems. For classical close-packing generates uniformity, while hierarchical arrangement yields heterogeneity, the asymmetrical relations which connect the levels looking after that. (Try to find the most rational way of packing equal balls on concentric spheres.)

Newton saw this, and seeking to account for the contrasted densities of bodies by assuming that they were in varying degrees porous, made an experiment from which something can still be learnt. In his mind's eye Newton arranged solid particles cohering in a stable pattern under mutual attractions and repulsions so that half the space was left void. Then he arranged these patterns similarly on a larger scale, again so as to leave the void at this second level also one half, and so on, until at the fifth level only 1/32 of the total space was filled with matter, while 31/32 was void, as he demonstrated to David Gregory in a conversation on December 21, 1705.[8]

This conjecture was worthy of its author. Some two centuries later Newton's idea, with appropriate modifications, would in effect provide a model for many kinds of heterogeneous systems, from the contrasted atoms of the Periodic Table, the

8. See p. 29, 21 December 1705, "D. Gregory, I. Newton and Their Circle." In *Extracts from Gregory's Memoranda 1677-1708*. ed. Hiscock, 1937.

hierarchy of electronic states in different atoms, the magic numbers of the atomic nuclei, and Pauling's recent speculative hierarchical model of the atomic nucleus. Differences between one thing and another, be they atomic or cosmic, invite the inference of hierarchical central arrangements of similar units.

Fifty years later the German astronomer, Lambert, following Newton (unawares?) and with the same aim of heterogeneity, but this time in the cosmos, tried a hierarchical arrangement, using a sequence of levels of gravitating orbital systems. Every system has a central mass and every orbiting unit is itself a smaller "solar system," and so on, like a Ptolemaic hierarchy of epicycles or a Calder mobile with a hierarchy of parts rotating in parallel planes. In 1761 this hierarchy of central suns was a brilliant idea, but wrong.

Then in 1907 the Irish-born physicist, Fournier d'Albe, published a diagram and a numerical description of an infinite hierarchical universe, designed to meet the Olbers radiative and Seeliger gravitational conditions, built on an octahedral principle from the smallest to infinitely large units. The six parts at each level are at the vertices of a regular octahedron, itself repeated six times at the next level, and so on, generating in this case a hierarchy of empty centres. This was not merely brilliant, but timely and fertile. For the Swedish astronomer, Charlier, a few weeks after reading Fournier d'Albe, generalized his octahedra into loose groupings, from single stars to what are now called galaxies, clusters, super-clusters, and so on – the first professionally designed infinite universe, for Fournier d'Albe was not an astronomer. This, like Newton's, was a model of non-uniform density.

Observe how the ground has been prepared for us by Newton, Lambert, Fournier d'Albe, and Charlier. So well indeed that certain working hypotheses are tempting, though they may only apply to restricted cases. (i) Structural hierarchies can give a rational quantitative representation of spatial *heterogeneity*. This is a heterogeneous universe; it is only homogeneous and isotropic to the zero-th order. (ii) Structural hierarchies are often *centrally* ordered at each level. Central forces may be, as it were, a theoretical bridge towards central ordering, since what counts is often a hierarchy of centres, occupied or not. (iii) Where a system is "sufficiently ordered" and "sufficiently nearly stationary" (terms to be clarified), three-dimensional *geometrical* relations (i.e., lengths or angles) may play a fundamental role. In sufficiently ordered quasi-stationary systems, mechanical and kinematic representations may for certain purposes be unnecessary. It is conceivable, in principle, that under certain conditions everything is derivable from angles. It seems that theory may sometimes pass rather easily from central geometrical hierarchical models to the heterogeneous properties of static, stationary, or near-equilibrium systems, thus opening the way towards a physics of hierarchy.

I suggest that the time has arrived for the gradual development of a comprehensive physical theory of the structural hierarchies of nature, perhaps a geometrical physics of clusters, nested structures, and nuclei of atoms and crystals, etc., including the processes which produce these. And of the biophysical hierarchies too, followed later by the social.

But this vista provokes a question. If structural hierarchies should tomorrow enjoy prestige in physics and biophysics, can we exact scientists trust others not to misuse the limited analogy to the human realm? I have no patience with such anxieties; fear is a bad guide. But let me state bluntly: spatial hierarchy does not imply anything homogeneous, monolithic, totalitarian, or derogatory to the human person himself, both as individual and as community, a concourse of hierarchies. I apologize for stating the obvious.

In any case it is too late to propose alternatives such as "holarchy" (Koestler 1967), for the term "hierarchy" began what I call its come-back, or entry into exact scientific thought, more than a hundred years ago. In 1854 Comte discussed the "hierarchy of the sciences," in 1858 Virchow considered the "hierarchy" of pathological conditions in medicine, in 1864 Bowen identified a "hierarchy of concepts" in science, and in 1881 Hughlings Jackson analyzed the "hierarchy of nervous centres" in man (though here the use of the term may have different implications). Though he did not use the term "hierarchy," A. Cayley's papers[9] (1857-1875) on the analytical forms which he called "Trees" are relevant. By the turn of the century both the idea and the term were in frequent use, e.g., in taxonomy and in grading the forms of energy. Between 1900 and 1950 at least seventeen pioneers used the term, or the idea, in the analysis of physical, biological, or social systems. Here are the names known to me: ten in biology: Stern, Hartmann, Weiss, Bertalanffy, Woodger, Needham, Goldstein, Gerard, Novikoff, and myself; four in sociology: Emerson, Redfield, Kroeber, Allee; three in physics and cosmology: Fournier d'Albe, Charlier, Weyl. These lists are not exhaustive; there were surely more. What matters is not names and dates but the emphasis put on a term, its meaning, its timeliness, and the use to which it is put.

Between 1945 and 1955, owing mainly to the advance of molecular biology and of astronomy, the frontier of scientific thought moved rapidly forward and the presence of structural hierarchies in the universe, particularly in organisms and in the cosmos, became a commonplace to many of these concerned with structure. I can only mention a few of those who, since 1950, have published significant scientific studies: Quastler (1955), Bernal (1958, 1962), Beckner (1959), Platt (1961), Simon (1962), Prosser (1965), Casper and Klug (1966), mainly in molecular biology, and Weizsäcker (1951), Layzer (1964), and Wilson (1965) in cosmology.

Here a caution is necessary. Scientific opinion is never homogeneous, and it takes time for new ideas to become widely accepted. The fact that many scientists are now actively

9. "On the Analytical Forms Called Trees," *Brit. Assoc. Report*. (257) 1875. Also *Phil. Mag* XIII. (172) 1857 and XX (374) 1859.

studying structural hierarchies does not mean that the concept has already been established. Certainly not, for its clarification and justification in an authentic physical law or a new principle of ordering still lies ahead. But the realm is now wide open, and interesting possibilities are in sight. I will mention only one. Structural hierarchies are perhaps more easily understood in terms of discrete particles, rather than by using a field or continuum representation. For me the discrete levels of a hierarchy invite discrete assumptions. I suggest that continua will soon be recognized as mere residuals of what I will call 19th century *macrosocopy*.

I have said enough to suggest the timeliness of this symposium. What is less clear, indeed still rather dark, is the optimal path towards an authentic, physical, and biophysical theory of the inorganic and organic structural hierarchies of levels with their similiaities and differences. But it may be that this formulation fails to make the crucial point explicit, and that the real need is for a systematic and exhaustive survey of *the types of three-dimensional spatial ordering* which characterize the more important levels in both realms. In any case it may be hoped that this symposium will facilitate a convergence of specialists on the collective task of constructing a comprehensive theory of the structural hierarchies of the universe.

Certain more technical issues are better treated separately. Some of these are covered by the *Five Questions* of the Note which follows (Whyte, this volume).

REFERENCES

Allee, W. C. 1949. *Principles of Animal Ecology*. Philadelphia: Saunders.

Ando, A.; Fisher, F.; and Simon, H. A. 1963. *Essays on the Structure of Social Science Models*. Cambridge: Massachusetts Institute of Technology Press.

Beckner, M. 1959. *The Biological Way of Thought*. New York: Columbia Univeristy Press. Reprinted, 1968. Berkeley: University of California Press paperback.

Bernal, J. D. 1958. "Structure Arrangements of Macromolecules." *Faraday Soc. Discussions* 25:7-18.

——. 1962. "Structure of Molecules." Chapter 3 in *Comprehensive Biochemistry* (4 vols.), eds. Florkin and Stotz. New York: Elsevier.

Bertalanffy, L. von. 1952. *Problems of Life*. New York: John Wiley & Sons.

Bowen, F. 1864. *Treatise on Logic.* Cambridge, Mass.

Caspar, D. L. D., and Klug, A. 1963. "Structure and Assembly of Regular Virus Particles." In *Viruses, Nucleic Acids, and Cancer*, pp. 27-39. Baltimore: Williams and Wilkins Co.

Charlier, C. V. L. 1908. "Wie eine unendlich Welt aufgebaut sein kann." *Arkiv för Matematik, Astronomi och Fysik,* Band 4, No. 24.

——. 1922. "How an Infinite World May Be Built Up." *Arkiv för Matematik, Astronomi och Fysik,* Band 16, No. 22, pp. 1-34.

Emerson, A. E. 1954. "Dynamic Homeastasis." *Sci. Monthly* 78:67-85.

Fournier d'Albe, E. E. 1907. *Two New Worlds*. London.

Gerard, R. W. 1940. "Organism, Society and Science." *Scientific Monthly* 50:340-50; 403-12; 530-35.

Goldschmidt, R. B. 1955. *Theoretical Genetics*. Berkeley: University of California Press.

Goldstein, Kurt. 1939. *The Organism*. New York: American Book Co. Reprinted, 1963. Boston: Beacon Press paperback.

Grene, M. 1967. "Biology and the Problem of Levels of Reality." *New Scholasticism* 41:427-49.

Harris, E. E. 1965. "Wholeness and Hierarchy." Chapter 7 in *Foundations of Metaphysics in Science*. New York: Humanities.

Jackson, J. Hughlings. 1958. *Selected Writings* (2 vols.), ed. J. Taylor. New York: Basic Books.

Koestler, A. 1967. *The Ghost in the machine.* New York: Macmillan.

Koestler, A. and Smythies, eds. (In press.) *Beyond Reductionism.* London: Hutchinson (edited proceedings of conference on hierarchy, Austria, August 1968).

Kroeber, A. O. 1949. "Concept of Culture in Science." *J. Gen. Educ.* 3:182-196.

Lambert, J. H. 1761. *Kosmologische Brieffe.* Leipzig. Translated, 1800, by James Jacque, *System of the World.* London: Vernon and Hood.

Layzer, D. 1964. "The Formation of Starts and Galaxies: Unified Hypothesis." In *Annual Review of Astronomy and Astrophysics,* Vol. 2, eds. Goldberg, Deutsch, Layzer, pp. 341-62. Palo Alto, Calif.: Annual Reviews, Inc.

Lehninger, A. L. 1965. *Bioenergetics.* New York: W. A. Benjamin Publishing Co.

Lindenmayer, A. 1964. "Life Cycles as Hierarchical Relations." In *Form and Strategy in Science*, eds. J. R. Gregg and F. T. C. Harris. Dordrecht: D. Reidel Publishing Co.

Lovejoy, A. O. 1936. *The Great Chain of Being.* Cambridge: Harvard University Press. Reprinted, 1960. New York: Harper Torchbook.

Needham, J. 1943. "Integrative Levels: A Revaluation of the Idea of Progress." In *Time: The Refreshing River*, pp. 233-72. London: George Allen & Unwin.

——. 1945. "A Note on Dr. Novikoff's Article." *Science* 101:582.

Novikoff, Alex B. 1945. "The Concept of Integrative Levels and Biology." *Science* 101:209-15.

Platt, John R. 1961. "Properties of Large Molecules That Go Beyond the Properties of Their Chemical Sub-Groups." *J. Theoret. Biol.* 1:342-58.

Prosser, C. Ladd. 1965. "Levels of Biological Organization and Their Physiological Significance." In *Ideas in Modern Biology*, ed. S. A. Moore, pp. 359-90. New York: Doubleday.

Quastler, H. 1955. "The Status of Information Theory in Biology." In *Symposium on Information Theory in Biology*, eds. Yockey, Platzman, and Quastler, pp. 399-401. New York: Pergamon.

Redfield, R. 1942. Introduction in *Levels of Integration in Biological and Social Systems*, ed. R. Redfield, pp. 5-26. Lancaster, Pa.: Jacques Catell Press.

Shapley, Harlow. 1930. *Flights from Chaos*. New York: McGraw-Hill.

Simon, H. A. 1962. "The Architecture of Complexity." *Proc. Amer. Philos. Soc.* 106:476-82.

Smith, Cyril S. 1965. "Structure, Substructure, Superstructure." In *Structure in Art and in Science*, ed. G. Kepes, pp. 29-41. New York: Braziller.

Virchow, R. 1858. *Die Cellularpathologie*. Berlin: Hirschwald.

Weiss, Paul. 1925. "Tierisches Verhalten als 'Systemreaktion'." *Biologia Generalis* 1:168-248.

——. 1967. "One Plus One Does Not Equal Two." In *The Neurosciences*, eds. Querton, Melnechuk, Schmitt, pp. 801-21. New York: Rockefeller University Press.

Weizsäcker, C. F. von. 1951. "The Evolution of Galaxies and Stars." *Astrophys. J.* 114:165-86.

Weyl, H. 1949. "Chemical Valence and the Hierarchy of Structures." Appendix D in *Philosophy of Mathematics and Natural Science*. Princeton, N.J.: Princeton University Press.

Whyte, L. L. 1949. *Unitary Principles in Physics and Biology*. New York: Holt.

——. 1965. *Internal Factors in Evolution*. New York: Braziller.

——. 1969. "Organic Structural Hierarchies." In *Unity and Diversity in Systems*. Essays in honour of L. von Bertalanffy, eds. R. G. Jones and G. Brandl. New York: Braziller.

Wilson, A. G. 1965. "Olbers' Paradox and Cosmology." The RAND Corp. Paper P-3256.

——. 1967. "Morphology and Modularity." In *New Methods of Thought and Procedure*, eds. Zwicky and Wilson, pp. 298-313. New York: Spinger-Verlag.

Woodger, J. H. 1929. *Biological Principles*. London: Routledge and Kegan Paul.

■

The Metaphysics, Epistemology and Methodology of Levels

Mario Bunge*

INTRODUCTION

Although this symposium is ostensibly devoted to hierarchies, not all the participants in it mean by 'hierarchy' a set partially ordered by an antisymmetric relation of domination or command but rather something less artificial and odious. I suspect that what most of us are concerned with is what biologists and social scientists have been calling *integrative levels,* or *levels of organization,* for the past half century or so. Be that as it may, I will start by elucidating the two concepts in question, namely those of hierarchy and level, and shall then proceed to state and discuss a few metaphysical, though not wild, theses concerning the multilevel structure of the world, as well as the consequences of these ideas for both epistemology (the general theory of knowledge) and the methodology of scientific research.

HIERARCHY

When a term is employed in a catch-all capacity, confusion is bound to arise and it will persist until conceptual analysis sets in. This is the case with the word 'hierarchy' in this symposium. If we wish to be understood and seek to minimize confusion, it will be necessary to begin by looking up the word in a dictionary and then try to give an exact elucidation of one or more of its meanings. Assuming – with no reason whatsoever – that the first task has been performed, we turn presently to the second one, i.e., to proposing an analysis attempting (but perhaps failing) to recapture the main traits of the general concept of hierarchy.

Strictly speaking, a hierarchy or hierarchical structure is a set equipped with a relation of domination or its converse, subordination. More precisely, the concept may be characterized by the following:[1]

DEFINITION 1. H *is a* hierarchy *if and only if it is an ordered triple* $H = \langle S, b, D \rangle$ *where* S *is a nonempty set,* b *a*

**Department of Philosophy, McGill University, Montreal.*

1. For a different though related definition of a hierarchy, see Woodger (1954).

distinguished element of S, and D a binary relation in S, such that

H1. *S has a single beginner, namely, b.* [That is, H has one and only one supreme commander.]

H2. *b stands in some power of D to every other member of S.* [That is, no matter how low in the hierarchy an element of S may stand, it still is under the command of the beginner.]

H3. *For any given element y of S except b, there is exactly one other element x of S such that Dxy.* [That is, every member has a single boss.]

H4. *D is antisymmetric and transitive.* [Togetherness but no back talking.]

H5. *D represents* [mirrors] *domination or power.* [That is, S is not merely a partially ordered set with a first element: the behavior of each element of S save its beginner is ultimately determined by its superiors.]

The five preceding statements constitute the axiomatic foundations of the microtheory of hierarchies. Any specific structure has to satisfy the above axioms if it is to qualify as a hierarchy: if it fails to satisfy them all then it is not a hierarchical structure – by definition of "hierarchy."

The preceding theory allows one to elucidate two further concepts, those of rank and superiority, in terms of concepts elucidated in that theory. The elucidation is effected by the following conventions.

DEFINITION 2. *A subset S_n of S, other than b, constitutes the n*th rank *of the hierarchy H if and only if, for every x in S_n, b holds the relation D^n to x.* [That is, the nth rank is n steps removed from the beginner.]

DEFINITION 3. *The rank S_n in H is* higher *than the rank S_n in H if and only if $n < n$.*

A diagram of a hierarchy is a finite tree branching out of a single point (namely *b*) and with no loops. The diagram is an open-ended directed graph of the kind studied in graph theory, a branch of combinatorial topology. But graph theory is concerned with graph structure only: it makes no place to our condition H5 above, which is a semantical not an algebraic clause. On the other hand graph theory is general enough to allow for an extension of our definition of a hierarchy to cover the case of many bosses and the possibility of multiple subordination – e.g., of every single member of *S* being dominated by two superiors.

It would seem that hierarchies are a human invention: in nature reciprocal action, rather than unidirectional action, seems to be the rule (Bunge 1959a). Hierarchical structures are found in society, e.g., in armies and in old-fashioned universities; but there are no clear cases of hierarchy in physics or in biology. True, a taxonomic system is a hierarchy (Woodger 1952, Bunge 1967), but then such a classification is not a concrete individual system but a system of concepts (sets) held together by the inclusion relation. Also, a body cell and its descendants may be regarded as a hierarchy (Lindenmayer 1964) – as long as the relation *D* in Definition 1 be not construed as domination but rather as "generates" or "produces" or "gives rise to." In short, one-sided domination seems to be an artifact. For this reason I submit that it is misleading to speak of hierarchies in nature, particularly when referring to evolutionary lines. The hierarchical universes imagined by Aristotle, Plotinus, and their Christian followers, were swept aside by the mechanistic science and metaphysics that emerged in the 17th century. In turn, the contemporary successor of the single level universe imagined by the mechanists is, I submit, a multilevel cosmos in which levels of organization have replaced hierarchical ranks. Let us then turn to an analysis of the level concept.

LEVEL

The term 'level' is highly ambiguous. It will therefore be convenient to state explicitly what we mean by it. The following definitions should serve this purpose.[2] The key concept in them is that of emergence or appearance of qualitative novelty in a process.

DEFINITION 4. *L is a* level structure *if and only if it is an ordered pair* $L = \langle S, E \rangle$ *where S is a family of sets of individual systems, E is a binary relation in S, such that*

L1. *Every member of S is a set of systems that are equivalent in some respect.* [That is, every member of the family *S* is a natural class – an equivalence class of systems sharing their basic properties and laws.]

L2. *E is a one-many, reflexive and transitive relation in S.*

L3. *E represents* [mirrors] *emergence or coming into being of novelty of qualitatively new systems in a process.*

DEFINITION 5. *A set of individual systems is* [constitutes] *a* level *if and only if it is a member of the family S of a level structure L.* [That is, a homogeneous collection of things constitutes a level provided neither *E* nor its converse hold among the members of the collection.]

DEFINITION 6. *A level is* newer *than another level of the same level structure if and only if the former has emerged from the latter.*

In less pompous terms, a level is an assembly of things of a definite kind, e.g., a collection of systems characterized by a definite set of properties and laws, and such that it belongs to an evolutionary line, though not necessarily to a line of biological descent. Some of the emergent characteristics or *nova* are the exclusive property of the given level. Furthermore, 'newer' does not mean "higher" or "superior" (a value concept) but just "later" in the game. For every *novum* has presumably emerged, in the course of a process, from preexisting levels. Likewise, in any evolutionary process some properties are lost.

2. For a different though related characterization see Bunge (1959b, ch. 5 and 1963, ch. 3).

Familiar examples of level structures are:

- Elementary particles – atomic nuclei – atoms – molecules – bodies.
- Physical systems – chemical systems – organisms – ecosystems.
- Physical processes – chemical processes – biological processes – psysical processes – social processes (= human histories).
- Material production – social life – intellectual culture.
- Physics – Chemistry – biology – psychology–sociology–history.

None of the preceding level structures qualifies as a hierarchy, since the concept of domination, essential to hierarchy, is absent from the idea of level. So is the concept of superiority, although one often calls 'higher' the newer levels on the mistaken progressivist assumption that the newer the better. Not even the concept of order is involved in the one of level structure, since the relation E of emergence is not asymmetric: while some level structures are ordered sets others are not. Thus a phylogenetic ladder is a partially ordered set, but the following categorization of the furniture of the universe is not ordered. [Incidentally, it constitutes the merger of two level structures: nature and society. (Bunge 1960)]

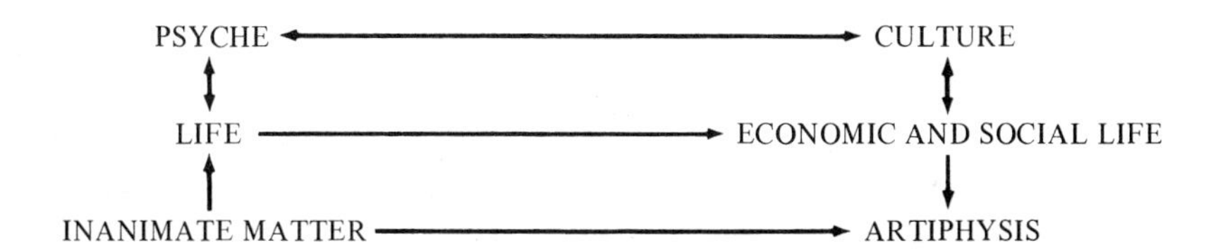

That this concept of level has been much alive in science since the days of Darwin is well-known. That it suggests a whole new metaphysics and a whole new epistemology, will be seen presently.

METAPHYSICS OF LEVELS

Consider the following metaphysical (ontological) hypotheses.

O1. *Reality (= the world) is a level structure such that every existent belongs to at least one level of that structure.* The thesis that all things group themselves naturally into sectors every one of which emerges out of previous sectors, is just a sweeping generalization of the theory of evolution. It competes with the Plotinian idea that the world is a hierarchy, a chain of being created once and for all. It also contradicts every shade of monism, for it asserts the originality, hence the irreducibility, of every new level. In particular, O1 goes against physicalism (mechanism, vulgar materialism), for which everything is physical, i.e., every item satisfies physical laws alone. But the main level hypothesis contradicts also idealism, for which there exists something (life, soul, spirit, or history) which is altogether independent from the physical level. The level hypothesis O1 contradicts unqualified pluralism no less than it does monism: indeed, though the world is not homogeneous, it is not divided into separate and isolated realms either. On the other hand O1 is consistent with *integrated pluralism*, an ontology that proclaims both the diversity and the unity of the world.

A further thesis of integrated pluralism is

O2. *In the course of every emergence process (self-assembly or evolution) some properties, hence also some laws, are gained while others are lost.* In other words, emergence, though "creative" of new patterns, is not cumulative. Thus consciousness appears with the higher mammals, which on the other hand are restricted to narrower temperature intervals than many of their ancestors. Progress, then, when real at all is not uniform but partial, and in any case it is not inevitable.

O3. *The newer levels depend on the older ones both for their emergence and for their continued existence.* In other words, new levels do not spring out of the blue and, once emerged, they do not subsist in mid-air. The older levels support the newer, without necessarily tyrannizing them. Moreover this dependence is primarily but not totally unilateral, as the new levels may exert a secondary reaction on the older ones. Thus physicists can create new particles and engineers can harness some forces of nature while politicians can wipe out whole human races.

O4. *Every level has, within bounds, some autonomy and stability.* That is, the dependence of a given level on its supporting level(s) makes room for some play. Thus the biological properties are hardly altered if the isotopic composition of the constituent atoms changes.

O5. *Every event is primarily determined in accordance with the set of specific laws that characterize its own level(s) and the contiguous levels.* In other words, not every level takes full part in the determination of any given event: interconnections are usually restricted to one level or its neighbors. Thus economic events, though based on the laws of life and the mind, obey economic laws.

Before passing judgment on these metaphysical hypotheses let us find out what they imply.

EPISTEMOLOGY OF LEVELS

If reality is organized into levels and is knowable, then that level structure must somehow be mapped in our scientific knowledge. That is, the ontological theses concerning levels, if true, must have epistemological partners that can be checked by examining knowledge itself. This does not mean that epistemology will mirror ontology in a simple way: the correspondence may be a very complex one. This is, in fact, what the following epistemological principles suggest.

E1. *The real level structure is knowable and scientific knowledge is a level structure that matches the former.* Stated negatively: no single science embraces the whole of reality. This thesis contradicts reductionism, the epistemological partner of monism.

E2. *Every newly formed science has its peculiar objects and special methods. And, although every science retains some of the ideas typical of the parent science(s), it does not preserve them all and it introduces new concepts absent from the latter.* Equivalently: even though all sciences share a general method (the scientific method) and a goal (the discovery of laws), every particular science deals in its peculiar way with a given aspect of the world. The unity of science lies in its basic language (logic and mathematics) and its general method rather than in a single object.

E3. *The understanding of any level is greatly deepened by research into the adjacent levels, particularly the underlying ones.* That is, every level of science (physics, biology, psychology, sociology) logically presupposes, for its advance in depth, the level(s) dealing with the corresponding older levels of reality. However, the levels of knowledge have not emerged in a single evolutionary process but have often had independent origins and developments. In other words, the level structure we call 'scientific knowledge' need not evolve the same way the world does: the various chapters of scientific research are apt to originate in a rather capricious way and to grow (or decay) at different rates.

E4. *Every level of science has, within bounds, some autonomy and stability.* For example, not every revolution in physics has had an impact on biology. And many psychological laws can be found without digging into their neurological bases. This is not only because we ignore most interlevel relations but also because of the objective comparative stability of real levels.

E5. *Every system and every event can be accounted for (described, explained or predicted, as the case may be) primarily in terms of its own levels and the adjoining levels, without necessarily involving the whole level structure.* That is, the knowledge immediately relevant to the level concerned should be scanned and reinforced before jumping to another level. For example, most historical events can be accounted for without resorting to physics and chemistry, but they cannot be properly understood without some behavioral science (including economics), which in turn calls for some biology.

The preceding bunch of epistemological theses outlines the impact of the assumed level structure of the world on our knowledge of it. Let us now proceed to inquire into the consequences of our ontological and epistemological hypotheses for methodology.

METHODOLOGY OF LEVELS

If there are levels in reality and in our knowledge of it, then surely they must leave their imprint on our behavior in the search for knowledge. That is, if there are levels then there must be some rules of scientific method that involve the idea of level and presuppose (and in turn confirm or infirm) the ontological and epistemological hypotheses about levels. In point of fact such rules can be discerned in actual research. Here go some of them.[3]

M1. *Start by limiting your inquiry to one level. Should this level prove insufficient, scratch its surface in search for further levels.* That is, stick to (methodological) reductionism till it fails: a reduction programme is as instructive when failing as when it succeeds. Methodological reductionism, if open-minded, is consistent with integrated pluralism; level separatism, on the other hand, is not.

M2. *Face emergence and try to explain it: begin by attempting to explain novelty away but, should this move fail, take it seriously.* That is, start by attempting to explain the new in terms of the old. If this strategy does not succeed, meet the challenge: take the *nova* by their horns. Do not ignore emergence and do not regard it as beyond comprehension either, the way emergentists (e.g., Alexander) and intuitionists (e.g., Bergson) used to do.

M3. *Explain the emergence of every level in terms of some of the older levels without skipping any intermediate level.* In other words, in order to dismount the emergence mechanism look at the older levels. For example, in order to explain the origin of organisms investigate the self-assembly of larger and larger molecules. Besides, when investigating interlevel relations try not to miss the intermediate levels and sublevels, if any. For

3. For a different presentation in the context of an analysis of the philosophical hypotheses that underlie scientific research, see Bunge (1967, ch. 5).

example, regard psychophysical laws – which are patterns not involving neurophysiological variables – as only temporary acquisitions to be completed later on – as end links of a chain to be filled in.

M4. *Begin by investigating your class of facts on their own level(s): introduce further levels only as required.* That is, be initially parsimonious in invoking further levels, for they may not intervene in any important way, and too many details clutter the way to theoretical models. But should this economy prove sterile, do not hesitate to bring in further levels. For example, do not populate society with more independent and rational minds than are needed to formulate the laws of economics, but do not neglect biology and psychology when it comes to explaining those laws.

M5. *Start by finding or applying the intralevel laws. Should this strategy fail, resort to hypothesizing or applying interlevel laws.*[4] For example, do not waste your time trying to account for the evolution of species in quantum-mechanical terms; but do use chemistry to explain the mutations that supply the raw material for selection.

Let us finally inquire into the logical relations among these norms and the previous ontological and epistemological hypotheses. To this end we shall assume that every norm of the form 'Do *x*' has been properly translated into the corresponding declarative sentence '*x* is done in order to get *y*,' where *x* is a name-holder for some means, and *y* a name-holder for some goal of scientific research.

CONCLUSION: METAPHYSICA A NOVO VINDICATA

We have formulated, discussed and illustrated – alas in a hurried way – three sets of principles concerning levels: a bunch O of ontological or metaphysical hypotheses concerning the level structure of the world, a collection E of epistemological theses about our knowledge of that structure, and a batch M of methodological principles involving levels of being and levels of knowing. These three sets are mutually related. Clearly, every member M_i of M suggests the corresponding component E_i of E,

4. The concepts of intralevel and interlevel laws have been analyzed and illustrated in Bunge (1967, ch. 6).

which in turn suggests the corresponding element O_i of O. These heuristic relations hold because the converse logical relations of implication may be assumed to hold among the corresponding principles – much as causes imply symptoms that suggest causes. In short we have, for every i from 1 to 5,

$$\text{If } O_i, \text{ then } E_i \tag{1}$$

and

$$\text{If } O_i \text{ and } E_i, \text{ then } M_i \tag{2}$$

(Obviously the converse conditions do not hold. Thus the mere diversity of the sciences could be a result of short-sightedness and convenience rather than an indication of a real level structure.)

Now, if our metaphysics is to be consistent with science and helpful to scientific research, then we cannot claim for the former more plausibility than the vicarious one it can derive from methodology. That is, metaphysics must prove its worth in science rather than as an independent enterprise. But even assuming that our methodology is correct, i.e., that it is actually employed in successful scientific research, we would not be justified in inferring that it verifies the metaphysics and the epistemology behind that methodology. Such an inference would be fallacious, this fallacy is so blatant and yet so common that it has got a standard name: it is called the fallacy of affirming the consequent. (Example: If positivism is right, then black-boxism, e.g., behaviorism, must be successful. Now, black-boxism is successful – though admittedly narrow – hence positivism must be right.)

All we can do is to draw the following *weak* (nondeductive) inferences:

If O_i, then E_i
Now, E_i (3)
Hence, maybe O_i

and

If O_i and E_i, then M_i
Now, M_i (4)
Hence, maybe O_i and E_i

In short, we can conclude that, to the extent to which our methodology works, the ontology and the epistemology behind it look *plausible*. And we may add that, whether or not our metaphysics and epistemology of levels are actually true, they seem to have been *fruitful*. They have certainly proved as fruitful, in recent times, as mechanistic reductionism was in its own time. Which is more than can be said, at the present time, for either monism and its epistemological partner, reductionism, or for pluralism and its associate, namely level separatism. In any case, right or wrong, integrated pluralism is worth being further explored as a candidate to the metaphysics of a science that acknowledges the existence of distinct but interrelated levels of organization.

ACKNOWLEDGMENTS

This paper has been supported by the Canada Council research grant 69-0300.

REFERENCES

Bunge, M. 1959a. *Causality*. Cambridge, Mass.: Harvard University Press.

——. 1959b. *Metascientific Queries*. Springfield, Ill.: Charles C. Thomas Publishing Co.

——. 1960. "On the Connections Among Levels." *Proceedings of the XIIth International Congress of Philosophy, Vol VI*. Florence: Sansoni.

——. 1963. *The Myth of Simplicity*. Englewood Cliffs, N.J.: Prentice-Hall.

——. 1967. *Scientific Research,* Vol I. New York: Springer-Verlag.

Lindenmayer, A. 1964. "Life Cycles as Hierarchical Relations." In *Form and Strategy in Science,* eds. J. R. Gregg and F. T. C. Harris. Dordrecht: D. Reidel Publishing Co.

Woodger, J. H. 1952. *Biology and Language*. Cambridge: Cambridge University Press.

——. 1954. "Problems Arising from the Application of Mathematical Logic to Biology." *Actes du 2ème Colloque International de Logique Mathèmatique*. Paris: Gauthier-Villars.

■

Foundations for a Scientific Theory of Hierarchical Systems

M. D. Mesarović and D. Macko*

From the discussions of hierarchies heard so far at this symposium, three points are emerging: (i) The term "hierarchy" is used to cover a variety of related yet distinct notions. (ii) The concepts of a hierarchy supposedly transcend various fields in the sense that they pertain to structural interrelationships rather ✓ than the specifics of the phenomena under consideration. (iii) The concepts of a hierarchy are presented in a fairly general and, one might even say, vague way, so that it is difficult to ✓ scientifically use such concepts. Since one does not have well defined problems or hypotheses which can be solved theoretically or tested experimentally, there is a need for a formal definition, i.e., a mathematical formalization of the various concepts of hierarchy.

The objective of our presentation is (i) to define more precisely three basic, yet distinct, concepts of a hierarchy: strata, layers and organizational levels and (ii) to point out a specific problem, namely, that of coordination as a characteristic problem of hierarchical systems. Only then can progress be made in understanding hierarchical systems. This is supported by historical evidence. What we know today about hierarchies is not more than we knew ten years ago. Even more, we are often reminded that the concepts of hierarchy were well appreciated in the past by ancient philosophers. Instead of being surprised at how much was known in the past, we could just as well contemplate on how little is known today. The theory of hierarchical systems and coordination we are presenting here (Mesarović, Macko and Takahara, 1968, 1969 and in press) hopefully will provide a basis for using hierarchical notions as a working hypothesis in scientific inquiry. Other relevant discussions can be found in Macko (1967); Mesarović, Sanders and Sprague (1964); and Mesarović (1966).

**Systems Research Center, Case Western University, Cleveland, Ohio, 44106.*

STRATA – LEVELS OF DESCRIPTION OR ABSTRACTION

Truly complex systems almost by definition evade complete and detailed descriptions. The dilemma in description is basically between simplicity, one of the prerequisites for understanding, and the need to take into account a complex system's numerous behavioral aspects. A resolution of this dilemma is sought through a hierarchical description. One describes the system by a family of models each concerned with the behavior of the system as viewed from a different level of abstraction. For each level there is a set of relevant features, variables, laws and principles in terms of which the system's behavior is described. For such a hierarchy to be effective it is necessary that the description on any level be considered independent of the description on other levels. Closure and frequency independence may make this possible. To distinguish this level-independent description concept of hierarchy from others we shall use the term *stratified system or stratified description*. The levels of abstraction involved in a stratified description will be referred to as *strata*.

Examples of stratified descriptions in the natural sciences are abundant. One can study living organisms on molecular, cellular, organ and total organism levels. Although a complete decoupling of levels cannot be fully justified, the assumed decoupling of levels enables the study of system behavior on any *stratum* to be performed in considerable detail. Neglecting the cross-strata interdependence leads to an incomplete understanding of the whole system behavior. Indeed, a restriction to biological-type inquires represents in itself an isolation since apparently the system under consideration can also be described on the stratum of conncern to chemistry or physics on one hand and ecology and perhaps social science on another.

One can take an electronic computer as an example of a stratified description of a man-made system. Its functioning is customarily described at least on two strata: in terms of physical laws describing the functioning and interconnection of the constituent parts and in terms of processing abstract nonphysical entities, such as digits or information sequences. On the stratum of information-processing one is concerned with such problems as computation, programming, etc., and the underlying physical basis of operation is not explicitly considered. Of course, a description of the system or some of its subsystems on other strata may also be of interest; the stratum of atomic physics for some component design or the so-called systems stratum for problems such as time sharing.

Other man-made stratified systems can be found in industrial automation. A completely automated industrial plant may be viewed on three strata: the physical processing of material and energy; the control and processing of information; and the economic operation in terms of efficiency and profit. A hierarchically integrated, computer controlled steel mill plant is diagrammed in Figure 1. Notice that on any of the three strata,

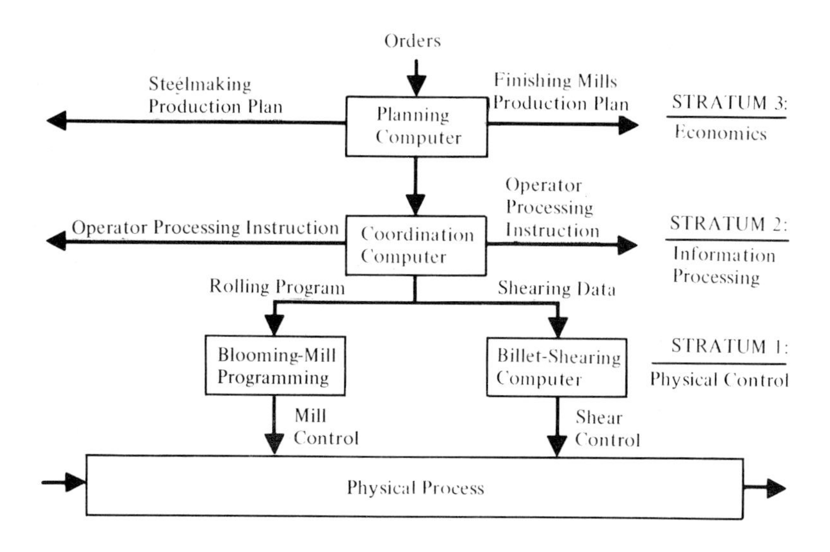

Figure 1 Strata in Automated Industrial Plant Operation

one is dealing with one and the same item—the finished product; on the first stratum it is viewed as a physical object to be changed in accordance with physical laws, on the second stratum it is viewed as a variable to be controlled and manipulated and on the third stratum it is viewed as an economic commodity. For each of these views of the system there is a different description, a different model; however, the system is of course one and the same.

How a single system can be described by a hierarchy of models is nicely illustrated by a system which produces a spoken literary composition (Polanyi 1968). There is only one output of the system: the actual, physical, utterance of a literary text. The operation of this system can be viewed from at least four strata: (i) The first stratum entails the generation of letters and describes the system as a sound-producing machine. (ii) The second stratum deals with the composition of letters into sequences that are acceptable as words in the grammar of a given language: the system is a word-producing machine. (iii) On the third stratum the system is considered in reference to the construction of sentences to express certain statements according to the given syntax and semantic rules. (iv) And on the fourth stratum the system is assessed by some literary aesthetic standards of the quality and literary value of the composition.

Other examples of stratified systems can easily be given. These examples suffice however to illustrate some general characteristics of a stratified description of a system.

1) *Selection of strata in which a given system is described depends upon the observer, his knowledge and interest in the operation of the system, although for many systems there are some strata which appear as natural or inherent.* In general, stratification is a matter of *interpretation* of the systems operation. If, for example, one is not familiar with the purpose of the electronic computer as a calculating machine, he is restricted to the stratum of physical laws. He can develop (given

sufficient time) a very detailed and accurate description of the system without being aware of the calculating aspect of the machine. Conversely, one can have an information-processing model without any knowledge of the physical laws involved. If one does not know the language used by the text-producing system, the best one can do is to recognize the sound-producing stratum. If one knows the language but has no literary standards nor background to evaluate the meaning and composition of the text, the higher strata are lost to him. The context in which the system is observed and used determines the fundamental strata.

2) *Contexts are not in general mutually related, and the principles or laws used to characterize the system on any stratum cannot in general be derived from the principles used on other strata.* For example, the principles of computation or programming are not derivable from the physical laws that govern the behavior of the computer on the lower strata and vice versa. Similarly, the grammar and syntax are not derivable from the physical laws of sound generation nor are the rules for a literary composition derivable (solely) from the grammar and syntax.

3) *There exists an asymmetrical interdependence between functioning of a system on different strata.* For a proper functioning of the system on a given stratum all the strata below have to function correctly. Therefore the requirements for proper functioning of the system on any stratum appear as conditions or constraints on the operations on the lower strata (Polanyi 1968). For example, in order for a computer to perform calculations, the physical process has to evolve in a fashion suitable to carry out the desired arithmetic operation. The generation of a given text depends on the operation of the system on the sound-producing stratum, and its production of sounds is determined by the requirements of the higher strata.

4) *Each stratum has its own set of terms, concepts and principles and what is considered as a system and its objects are different on each stratum. Furthermore, there is a hierarchy of*

objects and languages in which they are described. As a rule, the description on any stratum is less detailed than on the lower strata: an object on a given stratum becomes a relation on a lower stratum and an element becomes a set. A subsystem on a given stratum is a system on the stratum below. This relationship between strata is schematically shown in Figure 2.

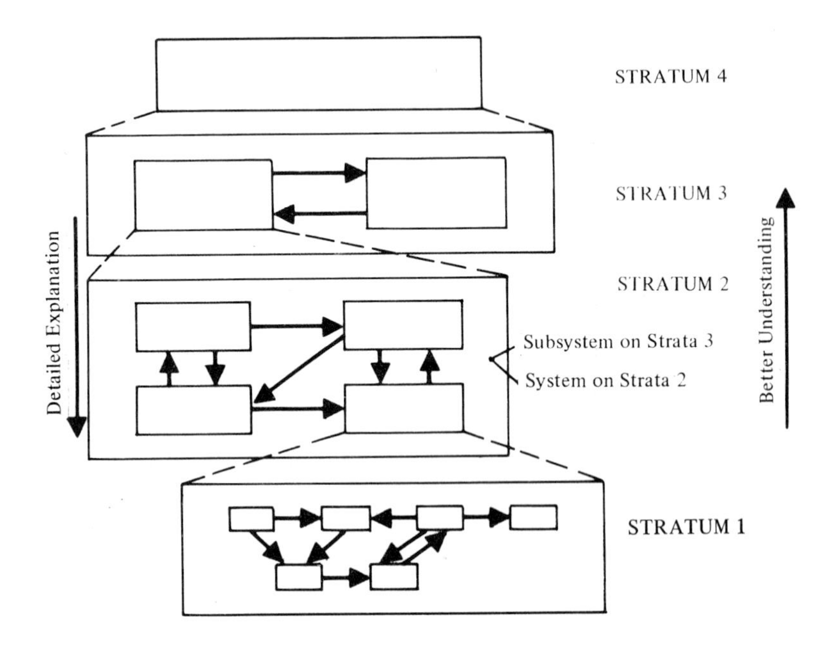

Figure 2 Multi-Strata System

On any given stratum, the behavior of the corresponding systems are studied in terms of their internal operation and evolution, while the question of how these systems interact so as to form a higher stratum system is studied on that higher stratum. This is important since it indicates that the studies on the lower strata are not necessarily "better," more fundamental or basic than on higher strata. On the lower strata one concentrates on the operation of the subsystems, leaving the study of their interrelationships for the higher strata. This would not be so if on the lower strata one would consider the entire system just as on the higher strata; this however is usually not the case since the principles and methodology on any stratum in general are not suitable for this procedure. It should be noticed that this object-system relationship between descriptions on various strata leads to a hierarchy of appropriate description languages. Since for each stratum there is given a different set of concepts and terms to be used for the description of the system on that stratum, there exists in general a different language.

5) *Starting from any given stratum, understanding of a system increases by crossing strata: moving down the hierarchy one obtains a more detailed explanation while moving up the hierarchy one obtains a deeper understanding of its significance.* It can be argued that the explanation in terms of the elements on the same strata is merely a description. For deeper understanding, the description should be given in terms of the elements on lower, more detailed, strata. As Bradley (1968) has pointed out "It is a serial process: The biologist explains transmission of heredity in terms of DNA replication; the biochemist explains the replication in terms of the formation of complementary nucleotide base pairs; the chemist explains base pairing in terms of hydrogen bonding; the molecular physicist explains hydrogen bonds in terms of intermolecular potential functions; the quantum mechanician explains potential functions in terms of the wave equation." By referring to lower strata one is able to explain more precisely and in more detail how the system functions to carry on a certain operation rather than which principles determine the particular operation to be performed. By moving up the hierarchy, the description becomes broader and refers to larger subsystems and longer periods of time. The analysis of a literary composition, starting from the alphabet, becomes richer as one comprehends and uses the principles of higher strata: grammar, syntax, rules of style and literally composition.

It would appear that for a proper understanding of complex systems the hierarchical approach is quite fundamental. Initially, one can confine his interest to one stratum depending upon the interest and experience and then increase understanding of its significance or explanation by moving up and down the hierarchy. Selection of the initial stratum is also affected by the simplicity of the description on that stratum.

LAYERS–LEVELS OF DECISION COMPLEXITY

Another concept of hierarchy appears in the context of a complex decision-making process. There are two trivial but important features of almost any real-life decision-making situation: to avoid decision by default, there is both a need to act without delay and an equally great need to take time to understand the situation better. In complex decision-making situations, the resolution of this dilemma is sought in a hierarchical approach. Essentially one defines a family of decision problems whose solution is attempted in a sequential manner. The solution of any problem in the sequence determines and fixes some parameters in the next problem so that the latter is completely specified and its solution can be attempted. In such a way the solution of a complex decision problem is substituted by the solution of a family of sequentially arranged simpler subproblems so that the solution of all the subproblems in the family implies the solution of the original problem. Such a hierarchy is termed a *hierarchy of decision layers,* and the entire decision-making system is referred to as a *multi-layer (decision) system.*

It is easy to give examples from everyday life of complex decision situations which are approached in a multi-layer fashion. Indeed, personal goals or objectives are as a rule vague and have to be translated into what might be called operational objectives which then provide a basis for the selection of a concrete course of action. For example, the goal of a young man might be to achieve happiness, but that vague goal has to be translated into objectives which lead to some specific actions: an objective has to be selected (say, to attend college, go into business, get married, etc.) which in turn leads to subgoals (say, to select a college, a major field of study, etc.).

Very often only after a subgoal is achieved is one in the position to evaluate whether the original goal is approached.

Consider now two examples of automated, man-made, decision systems (one from the field of artificial intelligence and the second from the industrial complex) in which the hierarchy appears more explicitly. In the heuristic programming approach to theorem-proving by computer proposed by Newell, Shaw and Simon (1959), the process of proving a theorem in a specified branch of mathematics consists of the following: The statement of the theorem is represented as an equality between two mathematical expressions. The expressions can be changed by applying transformations from a given set of "legal" or allowable transformations. Proving the theorem consists of changing the expressions on both sides of the equality until they become the same; in other words, the equality is transformed into identity. The process of transforming equality into identity is arranged hierarchically. For example in the propositional calculus a theorem is presented by the equality, say

$$R \wedge (\bar{P} \rightarrow Q) = (Q \vee P) \wedge R, \tag{1}$$

and the proof of the theorem consists of exhibiting a sequence of legal transformations that will show that the left and right sides can be transformed so as to be identical. This is attempted by a multi-layer system. The layers are defined in terms of differences which might exist between various expressions in the propositional calculus. Specifically, the following differences are recognized.

ΔV denotes that there exists a variable in one expression which does not appear in another;

ΔN denotes that a variable appears a different number of times;

ΔT denotes that the difference is in the negation of some variables;

ΔC denotes that different connectives are used;

ΔG denotes that the grouping is different;

ΔP denotes that the position of variables is different.

The differences are then ordered according to an assumed priority and assigned to separate decision layers starting at the top layer which is concerned with the most important difference. The task of the unit on each decision layer is to eliminate the respective difference. Each decision unit has a set of transformations which are deemed useful in eliminating the corresponding difference. The theorem-proving process consists then in presenting a theorem (1) to the top-most decision unit which after eliminating the respective difference presents the modified equality to the next unit. If all layers are successful, the equality is transformed into identity and the theorem is proven. It should be noticed that this discussion of the theorem-proving system indicates only the most essential structure of the theorem-proving system as proposed by Newell, Shaw and Simon (1959). The complete system is more complex: it provides for moving up and down the hierarchy to avoid being deadlocked if no solution is reached on any layer in a given time period. The equality can be fed back to one of the preceding layers or can temporarily be passed forward to the next layer with the provision that it will be returned to the higher layers if needed. At any rate, even this simplified description illustrates the need for multi-layer decision structure in complex decision situations.

Another example is provided by what is termed the functional hierarchy in decision problems under true uncertainties. This hierarchy emerges naturally in reference to three essential aspects of decision problems under uncertainty: (i) the selection of strategies to be used in the solution process; (ii) the reduction or elimination of uncertainties; and (iii) the search for a preferable or acceptable course of action under prespecified conditions. The functional hierarchy is shown in Figure 3. It contains three layers:

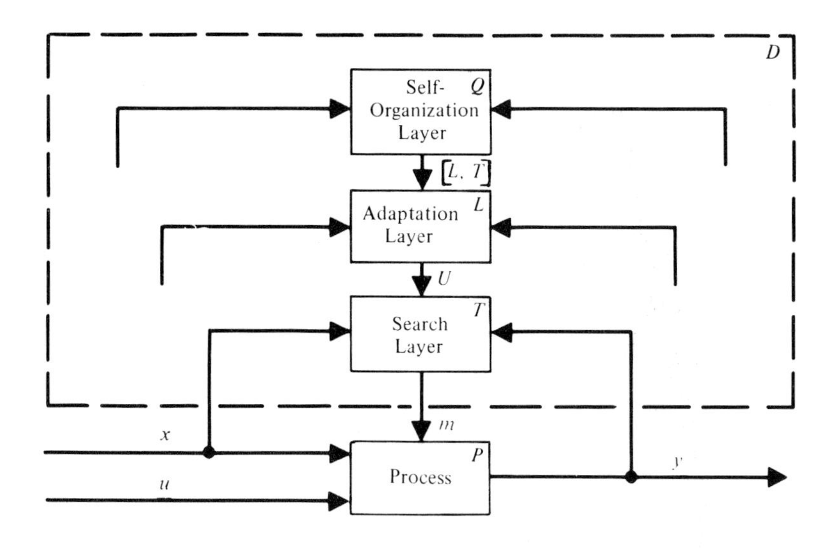

Figure 3 Functional Multi-Layer Decision Hierarchy

1) *Search layer:* The task of this layer is to select the course of action m. The decision unit on this layer accepts the outside data (information) and applies an algorithm (as specified by the higher layers) to derive a course of action. The algorithm might be defined directly as a solution map T giving a solution for any set of initial data or indirectly by means of a search process. For example, there is given an outcome function P and an evaluation function G and the selection of the action, say m, is based on the evaluation function G and in reference to P. Using set-theoretic (general systems theory) specification, the outcome function P is a mapping $P: M \times U \rightarrow Y$ where M is the set of alternative actions, Y is the set of outcomes while U is the set of uncertainties which can reflect in an equivalent fashion all the ignorance in relationship between the action m and the outcome y. Similarly, the evaluation function G is a mapping $G: M \times Y \rightarrow V$ where V is the set of values that can be associated with the performance of the system. If U is a unit set (i.e., it has a single element) or is void, which means that there are no uncertainties regarding the outcome for a given m, the selection can be based on optimization: find m in M such that

the value $v = G(m,P(m))$ is smaller than the value $v = G(m,P(m))$ for any other action m in M. If U is a large set, some other procedures have to be devised for the selection of the appropriate action, and other mappings in addition to P and G might have to be introduced. At any rate, to define the search problem for the first layer it is necessary to specify the uncertainty set U and the necessary relationships P, G, etc. These are provided by the units on the higher layers.

2) *Learning or Adaptation Layer:* The task of this layer is to specify the uncertainty set U used by the search layer. It should be noticed that the uncertainty set U is viewed here as encompassing all the ignorance about the behavior of the system and the environment and reflecting all the hypothesis about the possible source and types of uncertainties. U is derived, of course, on the basis of observations and communication. The inherent goal of the second-layer activity is to reduce the uncertainty set U. If the system and environment are stationary, the uncertainty set can be reduced to a unit set which corresponds to perfect learning as in controlled experimentation. However, it should be emphasized that U represents the uncertainties as assumed by the decision system rather than as they actually are. The second layer might very well need to change U altogether, increasing it if necessary and thereby acknowledging that some of the basic hypotheses were incorrect. Still, the overriding goal of the second or learning layer is to reduce the uncertainty set as much as possible and in this way simplify the job of the search layer.

3) *Self-organizing Layer:* This layer selects the structure, functions, and strategies which are used on the lower layers so that an overall goal (usually defined in terms which cannot easily be made operational) is being pursued as closely as possible. It can change the functions P and G on the first layer if the overall goal is not accomplished satisfactorily, or it can change the learning strategy used on the second layer if the estimation of uncertainties turns out not to be satisfactory.

Examples of the functional multi-layer hierarchy can be found in complex industrial control systems. The avid objective of introducing automation is to maximize profit, improve efficiency and minimize cost of operation. These stated economic objectives cannot be translated into concrete actions in the face of changing economic and technological conditions. In addition, the investment and operating cost of the automated system itself has to be taken into account as well as the technological constraints such as available hardware and prevailing engineering practice. All this leads to a hierarchical, multi-layer hierarchy in industrial systems shown in Figure 1.

It should be noted that the functional hierarchy as depicted in Figure 3 is based on the conceptual recognition of the essential functions in a complex decision system. It provides only a starting point for a rational approach to assign proper functions to different layers. In practice, a function on any one layer can be implemented by further decomposition. For example, in industrial automation the search layer is usually realized by a regulatory or direct control layer and optimization layer. The task of the regulatory layer is, in the face of inherent variations, to keep the respective variables near prespecified values which in turn are determined as so-called set-points by the optimization on the second layer.

ORGANIZATIONAL HIERARCHIES: MULTI-LEVEL MULTI-GOAL SYSTEMS

For this concept of hierarchy it is necessary that (i) the system consist of a family of interacting subsystems which are recognized explicitly, (ii) some of the subsystems are defined as decision (making) units and (iii) the decision units are arranged hierarchically in the sense that some of them are influenced or controlled by other decision units. This relationship between subsystems can be formalized and leads to a partial ordering relationship between subsystems.

These systems are denoted as multi-level multi-goal because, in general, various decision units comprising the system have conflicting goals. This conflict not only appears as the result of the evolution and composition of the system but also can be shown to be necessary (to a degree and in a given sense) for the efficient operation of the overall system.

The most prominent and generic examples of multi-level multi-goal hierarchical systems are formal human organizations. With this in mind, it would be hard to overemphasize the importance of these types of hierarchies. As Arrow (1964) has pointed out: "Among man's innovations the use of organization to accomplish his end is among both his greatest and his earliest. If we had no other evidence, we would know that complex organizations were necessary to the accomplishment of great construction tasks – planned cities like Nara or Kyoto, or monuments like Pyramids. . . . For less material ends we know of organization of the Inca Empire of Peru where a complex and far-flung state was administered in a highly systematic manner with a technology so poor as to include neither writing nor the wheel." Indeed many of our social problems can be traced to our lack of understanding of how to form or behave in an organizational context.

Many examples of multi-level multi-goal hierarchical systems can also be found in biology. It is legitimate then to ask, What is it that makes such a structure prevalent in the large scale systems area? A more detailed investigation (Mesarović, Macko and Takahara, in press) points out among others, the following reasons for the advantages offered by a multi-level hierarchical system:

1) Integration in a meaningful overall system of the already given family of independent but mutually interacting subsystems.

2) Utilization of units of limited physical size to accomplish the task which is beyond decision-making capacity of any individual unit.

3) Efficiency in utilization of the total resources; a given task might be accomplished with smaller expenditure in space or time if a multi-level approach is used.

4) Reliability and flexibility. The effect of a local change does not propagate so easily throughout the entire system.

One important characteristic of the multi-level multi-goal systems which sets them apart from conceptually simpler (although technically also quite complex) multi-variable decision systems should be emphasized. Namely, it is in the very nature of the multi-level multi-goal systems that the higher level units condition, yet do not completely control, the goal-seeking activities of the lower level units, The lower level decision units have to be given some freedom of action to select their own decision variables; these decisions might be, but are not necessarily, the ones which the higher level unit would select. Such a freedom of action is noticeable in any social or biological multi-level system. In man-made systems, a savings in the use of the resources available for decision-making can be achieved only if such a freedom of action is provided for at the lower levels. It can be shown that it is *essential for the effective usage of the multi-level structure that the decision units be given a freedom of action*; a suitable division of decision-making effort among the units on different levels should be established. Only then can the existence of hierarchy be justified.

This reasoning leads to a conceptually important classification of decision-making systems. Namely, with respect to the hierarchical arrangement of the decision units comprising a system, one can recognize the following categories of decision-making systems: *single-level single-goal systems; single-level multi-goal systems*; and *multi-level multi-goal systems*.

In the single-level single-goal class, a goal is defined for the overall system and all decision variables are selected so as to satisfy this goal. Technically, the solution of the decision problem that satisfies the overall goal can be quite complex since the problem is multi-variable and can involve optimization as well as prediction. Yet, the conceptual simplicity of the single-level single-goal system should be noticed – in particular, the absence of conflict within the boundaries of the system.

A system in the class of single-level multi-goal systems consists of a family of decision units each with its own goal. The goals of the system are not necessarily conflicting: a subfamily of decision units can form a coalition. There might exist a conflict between the decision units but none of them have the power to resolve the conflict.

Finally, the class of multi-level multi-goal systems is characterized by the existence of the hierarchical relation between the decision units of the systems. The existence of a supremal (top level) unit is the principal characteristic of this class of systems and the decision problem for the supremal unit is a principal problem specific for multi-level systems.

Having in mind the above classification, it becomes apparent that a new theory is needed to deal with the multi-level systems. It might be argued that control theory as currently conceived deals with single-level single-goal (although multi-variable and rather complex) decision problems and game theory and the theory of teams takes care of the single-level multi-goal systems. None of these theories are complete and much investigation is needed (i) to develop practical methods for the control of single-level single-goal systems or (ii) to understand the nature and effects of the conflict in single-level multi-goal systems. However, for at least certain classes of single-level systems the framework for developing a theory has already been established in the course of the extensive research over the past two or three decades. Apparently, a novel approach is needed to deal with the hierarchical, multilevel systems. A framework for the development of a mathematical theory for such systems is reported in Mesavorić, Macko and Takahara (in press).

COORDINATION, INTEGRATION AND CROSSING OF LEVELS

Progress in understanding hierarchies has to come from detailed and careful studies of some specific issues characteristic of hierarchical systems. Such a specific is that of coordination. A detailed mathematical theory of coordination has been developed and is reported in Mesarović, Macko and Takahara (in press). We shall limit our comments here to its implication for the broader issue of crossing levels in a hierarchy.

Consider an organizational-type hierarchy with two levels. There are n decision (making) units on the first level and one on the second level. The first-level decision units are interacting; the performance of one unit depends upon the action of other units on the same level. The actions of all first-level units are jointly applied to a so-called (controlled) process P or, alternatively, to mutually interacting local processes $P_1, \ldots, P_n$ which make up P. The second-level unit can affect the decision problems of the first-level units, its task being to influence the first-level units so that a goal valid for the organization as a whole is promoted. Since the first-level units are ultimately responsible for the actions (controls) applied to the process P the second-level unit has to influence them so that while pursuing its own goals it comes up with the decisions which are desirable for the overall higher level objective. The second-level unit achieves this by selecting variables which, like parameters, specify the first-level decision problems. These are termed intervention or coordination variables. More specifically, then, the decision problem of the ith first-level unit is parameterized by second-level intervention variables. There is in effect associated with the ith first-level unit, a family of decision problems, let them be denoted by $\Delta_i(\beta)$, $\beta \epsilon$ C, so that for any given intervention β the decision problem is fixed and hence the ith first-level unit can select the appropriate action.

The intervention variable is selected by the second-level unit as a solution of a decision problem. This decision problem is termed the coordination problem. It is essential for an understanding of the fundamental nature of a hierarchical structure and the potential advantage in such structures to realize the distinction which exists between the overall decision problem and the coordination problem. The overall problem is the problem which the entire organization is supposed to solve, indeed, it is in order to arrive at the solution of the overall problem that the organization exists. The coordination problem, on the other hand, is simply the decision problem of one of the units in the system, which because of its particular position is of special importance. However, the task of the second-level unit is not to solve the overall problem, but only to affect the first-level units so that the behavior of the overall system moves toward the solution of the overall problem.

To illustrate the distinction between the coordination problem and the overall problem consider a simple example. Let a two-level system have two decision units on the first level.

There are three kind of decision problems associated with the system: (i) the first-level units solve *the first-level problems* Δ_1 and Δ_2 as a means of arriving at their local action, which is applied to some process P, (ii) the second-level unit uses *the coordination problem* Δ_o in order to arrive at the intervention parameters necessary to coordinate the first-level units and, (iii) the second-level unit coordinator the first-level units in reference to an *overall problem* Δ defined for the system as a whole (the success of coordination is evaluated in terms of Δ). Let us define these types of problems for the example under consideration.

The first-level problem Δ_1 is defined by means of the model for the subprocess P_1 and a cost function G_1. Let the model for P_1 be given by the equation

$$y_1 = 2m_1 + u_1 \tag{2}$$

where m_1 is the local control, u_1 the input coming from the other subprocesses P_2, and y_1 is the local output. Let the cost function be the quadratic form

$$G_1(\beta, m_1, y_1) = m_1^2 + (y_1 - 1)^2 + \beta_1 u_1^2 - \beta_2 y_1^2 \quad (3)$$

Then, the problem Δ_1 is to find the values for the local control m_1 and the input u_1 which will minimize the cost as given by G_1 observing the constraint given by (2). This minimization of course is for a given β_1 and β_2 which are the intervention parameters specified by the coordinator.

The other first-level decision problem Δ_2 is defined quite analogously by means of a model for the local process P_2

$$y_2 = 2m_2 - u_2 \quad (4)$$

and a local cost function

$$G_2(\beta, m_2, y_2) = m_2^2 + (y_2 - 1)^2 - \beta_2 u_2^2 + \beta_2 y_2^2 \quad (5)$$

The problem Δ_2 is to find the values for the local control m_2 and the input u_2 which minimize G_2 observing the constraint (4) when given the interventions $\beta_1$and β_2.

Before proceeding with the formulation of the overall problem and the coordination problem, the following should be noted: the overall process which is under the control of the hierarchy is defined by a pair of equations whose simultaneous solution yields the outputs of the system, i.e.,

$$\begin{aligned} y_1 &= 2m_1 + y_2 \\ y_2 &= 2m_2 - y_1 \end{aligned} \quad (6)$$

Each of the first-level units is concerned only the the local processes and views the non-local variables as interactions, i.e., $u_1 = y_2$ and $u_2 = y_1$.

Because the outputs are the simultaneous solution of (6), the actual interactions are not known to the first-level units and this is precisely why their decision problems are defined so that u_1

and u_2 are treated as decision variables during the optimization process even though they are dependent variables.

To define the overall problem we use the sum of the local cost functions when $\beta_1 = \beta_2 = 0$

$$G(m,y) = m_1^2 + m_2^2 + (y_1 - 1)^2 + (y_2 - 1)^2. \tag{7}$$

Its solution is a pair (m_1, m_2) of controls which minimize (7) while observing the constraint (6).

To formulate the coordination problem let the pair $(\hat{m}_i(\beta), \hat{u}_i(\beta))$ for $i = 1,2$ denote the solution of the first level problem Δ_i for given $\beta = (\beta_1, \beta_2)$, while $u_1(m)$ and $u_2(m)$ denote the actual interactions when the control $m = (m_1, m_2)$, is applied to the process P defined by (6). The coordination problem is to find the values for the intervention parameters β_1 and β_2 so that the interaction inputs $\hat{u}_1(\beta)$ and $\hat{u}_2(\beta)$ as required by the local units are exactly those which will actually occur when $\hat{m}(\beta) = (\hat{m}_1(\beta), \hat{m}_2(\beta))$ is applied,

$$\begin{aligned} u_1(\beta) &= u_1(m(\beta)) \\ u_2(\beta) &= u_2(m(\beta)). \end{aligned} \tag{8}$$

In other words, a balance is sought between the desired and actual interactions in the sense that the actual interaction input to any local process will be that which the local unit needs to achieve its own local goal.

The entire process is then the following: Given the interventions β_1 and β_2, the first-level units select the pairs $(\hat{m}_1(\beta), \hat{u}_1(\beta))$ and $(\hat{m}_2(\beta), \hat{u}_2(\beta))$ respectively. The local controls $\hat{m}_1(\beta)$ and $\hat{m}_2(\beta)$ are applied to the process P. If the resulting interactions u_1 and u_2 are as required by the local units the overall goal is achieved, i.e., the selection of (β_1, β_1) is satisfactory. If not, the coordinator has to change β_1 and β_2 until such a condition is reached.

With the proceeding example in mind one can appreciate the following remarks:

(i) The task which the total organization is to perform is split among levels so that each unit is solving a subproblem but in such a way that the partial solutions add up to the solution of the original, overall task. It is this subdivision of labor which makes the hierarchical approach effective.

(ii) The units on the higher level are not concerned explicitly with the higher level overall goals but rather with different goals specific for their level of operation. The coordinator is concerned with balancing interactions. The higher level unit does not need a complete, detailed knowledge of the activities on the lower levels but rather is concerned only with some relevant feature such as the lack of interaction balance. The achievement of the overall goal, again, is implied if all the units on different levels function properly.

(iii) Explanation of the functioning of a hierarchical system should not be attempted in terms of the overall goal but rather in terms of the specific goals valid for each particular level. In this respect, a frequently made mistake is to assume the highest level unit is in charge of the overall goal. It is not! It has its own specific goal (that of coordination), and the overall goal is achieved only by the combined action of all units; it is implied if all the units function properly. This is the essence of the statement which we made at the beginning of this section, namely, integration of the overall system and harmonious behavior achieved by coordination appears to be one of the key concepts for the study of crossing of levels in a hierarchical system.

ACKNOWLEDGEMENTS

This research has been supported in part by Office of Naval Research Contract No. Nonr 1141(13).

REFERENCES

Arrow, Kenneth J. 1964. "Control in Large Organizations." *Management Sci.* 10:397-408.

Bradley, D. F. 1968. "Multilevel Systems and Biology." In *Systems Theory and Biology,* ed. M. D. Mesarović, pp. 35-58. New York: Springer-Verlag.

Macko, D. 1967. "General Systems Theory Approach to Multilevel Systems." Thesis: Case Western Reserve University.

Mesarović, M. D. 1966. "A Conceptual Framework for the Studies of Multilevel Multigoal Systems." Systems Research Center Report SRC 101-A-66-43.

Mesarović, M. D.; Sanders, J.; and Sprague, C. 1964. "An Approximate Approach to Organizations from a General Systems Viewpoint." In *New Perspectives in Organization Research,* ed. New York: John Wiley & Sons.

Mesarović, M. D.; Macko, D.; and Takahara, Y. 1968. "Structuring of Multilevel Systems." Proceedings of the IFAC Symposium, Düsseldorf.

——. 1969. "Two Coordination Principles and Their Application in Large Scale Systems Control." To be presented at IFAC Congress, Poland (June).

——. (in press). *Theory of Multi-Level, Hierarchical Systems.* New York: Academic Press.

Newell, A.; Shaw, J. C.; and Simon, H. A. 1959. "Report on a General Problem-Solving Program." *Information Processing.* Paris: UNESCO.

Polanyi, M. "Life's Irreducible Structure." *Science* 160:1308-1312.

■

Five Primary Questions

Lancelot Law Whyte*

I. Are those levels which are of most significance for physical and biophysical theory each distinguished by a particular type of three-dimensional spatial ordering of their first order parts? (Does this lead to a generalization of the concept of "phase" or homogeneous region in thermodynamics?)

II. If so, does this render possible a natural classification and enumeration of the most important levels in the inorganic, and in the organic structural hierarchies? (Does the enumeration of the levels in the organic hierarchy from atoms to living cells permit the determination of a finite number of crucial steps in the historical transition from inorganic materials to cellular organisms?)

III. What measurable properties (angles, lengths, times, masses, potentials, etc.) are associated with each level?

IV. If we define "morphic" as those processes "displaying a movement toward greater three-dimensional spatial order, symmetry, or form" (Whyte 1969), what is the relation of those morphic processes which generate levels (or structural units) of hierarchies to the entropic processes which tend to disperse them?

V. What is the relation of the hierarchy-generating morphic processes, and of the constraints imposed by the existence of a unit or level on the degrees of freedom of its parts, to known physical laws? (What cases of morphic processes, e.g., atomic nucleus formation, crystal nucleation, etc. are fully covered by Quantum Mechanics, Statistical Mechanics, or current Nuclear Theory, so that no empirical parameters must be assumed in addition? In what classes of cases do these theories fail to give a unique prescription for the type of representation to be used (i) for complex partly ordered systems or units, and (ii) for the processes by which they are formed?)

REFERENCES

Whyte, L. L. 1969. "Organic Structural Hierarchies." In *Unity and Diversity in Systems*, eds. R. G. Jones and G. Brandt. New York: Braziller.

■

**93 Redington Road, London, N.W. 3, England.*

Comments on the Use of the Term Hierarchy

Robert Rosen*

Biologists employ the term "hierarchy" in an inexact, non-etymological sense, to describe the way in which the functional activities or organisms are distributed into various identifiable levels. I tried to give an operational definition for this usage at the beginning of my paper (this volume). The question of how this usage is related to that described by Professor Bunge, which deals with discrete units related by a one-way "bossing" relation, is an interesting one. I would like to stress two aspects of this question.

First: in the biological situation, a particular biological function, at any level, tends to be *distributed* over much, if not all, of the entire system. Further, as I indicated in my paper, the whole sequence of levels is simultaneously superimposed on the same indivisible system. From this we conclude (a) that we cannot in general produce a set of discrete, disjoint units on which a "bossing" relation can be imposed (because of the distributed character of biological functions, which has as a consequence that any particular part of the system is simultaneously involved in a variety of different functional activities), and (b) since the recognition of levels in organisms operationally involves merely different descriptions of the activities of the *same* system, nothing that happens at any one level can be without consequences for the activities at all other levels. Thus, as Dr. Pattee has also stressed, the interaction between functional levels in a biological system is a reciprocal one, not unidirectional as a "bossing" relation must be. However, the question of whether a reciprocal relation can conveniently be decomposed and studied as a pair of "bossing" relations working in opposite directions is an interesting one, and should be studied further.

Second: the question of the origin of structure and organization in the universe, as studied by the cosmologist, seems very much analogous to the problem of how an initially homogeneous biological system (like a fertilized egg) or (a blastula of apparently identical cells) can spontaneously generate the complex patterns involved in the differentiation of such a system into an adult organism. A number of different models for the generation of such biological patterns have been proposed. These seem formally very similar to those described by the cosmologists at this symposium. Mathematically, these differentiation models all depend on the initial homogeneous state being one of unstable equilibrium, so that random disturbances will cause the system to move in

---

**Center for Theoretical Biology, State University of New York at Buffalo. Amherst, New York, 14226*

a determined manner toward a new state which must necessarily be inhomogeneous.[1]

Stated another way, in each of these models an initial random displacement acts autocatalytically, via a positive feedback loop, to move the system further away from homogeneity. It should be noted that these models also depend essentially on a nonlinearity somewhere in the system; either in the equations of motion themselves, or in the boundary conditions (where linear equations of motion break down as we approach limiting values of the state variables). The cosmological discussions I have heard, however, all seem to rely heavily on linearity, and a too heavy reliance on linearity would vitiate (in purely mathematical terms) any possibility of pattern generation of the kind envisioned by the cosmologist.

■

---

1. I've discussed some of these models and give further reference in *International Review of Cytology* 23:25-88 (1968) and *Bull. Math. Biophysics* 30:493-499 (1968).

Closure, Entity, and Level

Albert Wilson*

The manner of decomposition of a complex organism or structure into sub-components is arbitrary. With a scalpel in the dissecting room or with the knife of pure intellect, the decomposer has freedom to isolate many alternative sub-groupings. However, unless his knife follows the "natural interfaces," severing a minimum of connections in isolating the sub-components, his decomposition may prove to be confusing, uninteresting, and messy. Whereas all decompositions possess the kind of properties that are treated in classical set theory, those decompositions conforming to natural interfaces frequently reveal additional interesting properties. What we call the "natural interfaces" are identifiable either by the occurrence of a steep decrement in the number or strength of linkages crossing them, as developed by Simon (1962) in the concept of *near decomposibility,* or through the existence of some form of *closure.* The purpose of this note is to sketch how entity and level may be related to one or more forms of closure.

The most apparent from of closure is *topological* closure—the encompassing by (one or more) closed surfaces of a spatial neighborhood that coincides with or bounds the extension of a physical object. We thus perceive balls, donuts, strings, and sheets as topologically closed. In general, topological closure bestows finitude and convexity on objects and is a property of most entities that we differentiate by visual perception.

A second type of closure, associated with a neighborhood in time that coincides with or bounds the *duration* of an entity, may be called *temporal* closure. More abstract notions of closure may be employed to distinguish non-physical entities. Thus a *group* may be defined as a set of numbers, elements, or transformations that possess closure with respect to some *operation.* For example, the integers 0, 1, 2, 3, 4 form a group closed under addition modulo 5. This type of operational closure, when the number of elements is finite, joins temporal closure in being cyclical in the sense that some parameter follows a path that periodically returns to previously assumed values. Topological closure and cyclical closure can be related through various Fourier type transformations. Spatial representations (particles) and frequency representations (waves) may thus both be subsumed under the notion of closure. In addition isolation of entities may take the form of either physical separation or "detuning."

Not only may differentiatable entities and modules be described through the use of some form of closure or cyclical parameter, but many

**Douglas Advanced Research Laboratories, Huntington Beach, California, 92647*

notions of *level* may also be differentiated through closure. For example, levels in control hierarchies such as industrial corporations are determined by subsystems identifiable through various feed-back loops which are mappable onto a set of closed cyclical parameters. In modular hierarchies (Wilson 1967) levels and modules share a set of topological closures and when the modules are homogeneous the levels become identical to the modules.

The example of hierarchical cosmic sub-structures (Wilson 1969) shows that levels may be distinguished by a characteristic time or frequency, which is to say that each level is temporally closed. This suggests that the properties of space and time are closure properties of structures, bringing to mind the basic idea of Leibniz that space and time have no independent existence, but derive from the nature of structures. Einstein's equivalence of dynamics and geometry contained in his field equations (e.g., matter density determines spatial curvature) is also consistent with Leibniz's view and a departure from the Newtonian idea that all structure exists within an independent framework of space and time. It may then be that from the various closures and partial closures of structures and systems, we infer the descriptions we call space and time.

REFERENCES

Simon, H. A. 1962. "The Architecture of Complexity." *Proc. Amer. Philos. Soc.* 106:467-482.

Wilson, A. G. 1967. "Morphology and Modularity." In *New Methods of Thought and Procedure*, eds. Zwicky and Wilson. New York: Springer-Verlag.

——. 1969. "Hierarchical Structure in the Cosmos." (this volume).

■

Hierarchy: One Word, How Many Concepts?

Marjorie Grene*

To a philosophical observer, the first lesson of the conference was that the use of the word "hierarchy" is strikingly equivocal. In diverse disciplines, it signifies diverse concepts. Many of the speakers appeared to be addressing themselves to wholly disparate inquiries, guided by the word "hierarchy" and the special concept which it suggests in their technical usage. Coming to the problem of hierarchy from philosophical questions about living entities, I had the initial impression that astronomers use the term, in the main, to focus on a comparison of size in the aggregates they talk about. Perhaps the criterion of comparative density would be more appropriate; but an ordering by size was all I could get out of much of the astronomical talk on first hearing, and the film about the dimensions of celestial objects confirmed this impression. Biologists, on the other hand, when they worry about hierarchical organization in living things, are concerned about (at least) doubly determinate systems: systems such that an arrangement of the elements comprising the system constrains the behavior of the elements themselves, the controlling order thus constituting an upper level (not necessarily larger, however) in relation to the elements so ordered, which, in turn constitute a lower level of the system. Dr. Pattee's definition characterizes this kind of situation quite clearly. Artifact people, finally, are interested in hierarchies in a sense more like the biological than the astronomical concept.

This very rough division, however, fails to stand up to closer scrutiny. For one thing, there is after all something in common to both main uses of the term, something recognized by non-scientific common-sense as well as indeed "hierarchical": for in both cases the concept "hierarchy" is used as a guide to the *ranking* of some set of entities or operations. In both cases, moreover, it is allegedly *real* entities or processes that are so ranked. So the equivocation seems to be only partial after all; the concept of *control* is not always present, but *ranking* and a reference to *reality* are.

Further, even insofar as scientists from different disciplines are employing different concepts, the divisions are far from clear-cut and the meanings overlap. On the side of astronomy, for example, Dr. A. Wilson's elegant rescuscitation of the Leibnizian concept of space suggested that, in the most comprehensive way, it is the *ordering* of bodies in the universe that governs their distribution and their systematic behavior. Not only the structure and function of living things, therefore, but the universe itself, it would seem, may be the outcome of hierarchical relations – "hierarchical" now in something close to the biologists's or computer theorists's sense.

Department of Philosophy, University of California, Davis, California, 95616.

And on the other hand, there are other uses of the term in biology itself which seem to weaken its stricter meaning. Taxonomists mean by "hierarchies" simply a ranking of taxa such that each "higher" taxon includes those lower down the scale, from a given kingdom through a phylum. . . to a given species. There is here no vestige of "control" of one level by another, but only a logical arrangement of class concepts (Buck and Hull 1966). Indeed, it is by no means clear that the "higher" categories (such as "family," "order," "class," etc.) represent more than a human convenience for sorting out the many living individuals which exist and have existed. Thus even such visible aggregates as those ranked by astronomers are missing from the denotative import of the term in this case.[1]

What can a philosopher contribute to disentangling this confusing set of usages? Dr. Bunge exhorted us to eliminate the errant word from our vocabulary in all its looser meanings. Indeed, by this excision he would, it seems, eliminate *all* the uses actually made of the term by all the scientists concerned. For there is in none of the alleged hierarchies in question a "boss" in the original, etymological, priestly sense, which alone, for Dr. Bunge, makes a proper hierarchy. Such drastic revision of scientific – and philosophical – vocabularies, however, appears ill-advised. For one thing, philosophers cannot successfully prescribe to scientists the way they shall use their terms. If the astronomer and the biological or computer systems theorists do use a term in radically different ways, we just have to take note of this equivocation: it's like "sole" to eat and the "sole" of a shoe. Homonyms can be tolerated, so long as we recognize their existence and understand the term's several meainings, as Dr. Bunge himself has demonstrated, for example, in his precise analysis of alternate meanings of the term "level." (Bunge1959).

Yet that may not be the most reasonable solution in our case. There *is* something in common, as we have seen, in the various uses of the term: In every case there is *ranking* of some kind; in most cases, moreover, it is a ranking of *real* entities or of levels of organization, or levels of

1. Still another meaning from the side of biology was offered by Dr. Rosen, who understands by "hierarchy" any system which may be studied by alternative methods. Here, however, the one common criterion, that of some ranking operations, is missing; I believe it would be better, for the systems theorists's purposes, to adopt Dr. Pattee's definition, which permits us to see living systems as displaying what I have elsewhere called "ordinal" (rather than cardinal) causality (Grene 1966, Chp 8). It is of course true that one can study a hierarchical system either analytically (in terms of its lower level) or comprehensively (in terms of high level principles). But while the lower is necessary to the existence of the higher, the higher is not necessary to the lower; and while the higher constrains or controls the operation of the lower, the lower does not in this sense control the higher. This double asymmetry is missed by the reference to "alternative" methods alone.

performance, or real entities. Perhaps the meanings of "hierarchy" form a family in Wittgenstein's sense. Perhaps they even form a hierarchy (as suggested by Dr. A. Wilson in a recent conversation). Further clarification of this issue is plainly needed.

Nor, be it noted finally, is this merely a "semantic" issue. To clarify our concepts is to clarify our ways of apprenhending what there is. If a hierarchy of concepts of hierarchy turns out to be useful, and even necessary, for a series of scientific disciplines, it will be because we have discovered that the real processes and events studied by astronomers, biologists, computer scientists and epistemologists, themselves constitute hierarchies of hierarchies, in a fashion which, it is to be hoped, further cooperative ventures like this one may help to render more precise.

REFERENCES

Buck, R. C., and Hull, D. L. 1966. "The Logical Structure of the Linnaean Hierarchy." *Systematic Zoology* 15:97-111.

Bunge, M. 1959. "Do the Levels of Science Reflect the Levels of Being?" Chapter 5 in *Metascientific Queries.* Springfield, Illinois: Charles C Thomas.

Grene, M. 1966. *The Knower and the Known.* New York: Basic Books.

Part II

Inorganic Hierarchical Structures

Two primary hierarchies that occur in the inorganic world are the hierarchy whose primary bonding derives from electrical forces and whose levels are molecules, crystals, and crystalline aggregates, and the hierarchy whose primary bonding derives from gravitational force and whose levels are stars, galaxies and clusters of galaxies. In the first paper of Part II, Cyril Smith discusses the levels of organization in the first hierarchy – the super-atomic world. The existence of levels depends on repeatably local interactions and connections among which discontinuities eventually occur to give rise to larger groupings. But since each level is what the observer sees at certain resolutions, Smith considers that the structures that emerge on a larger scale may be partly illusory. An assembly of elements will not form a coherent aggregate unless the parts interact in such a way as to modify their internal structure and energy. The interfaces between entities at various levels may coincide with actual physical discontinuities or they may be only surfaces at which the gradient of some property changes sign. Smith concludes with six general principles that appear to hold for many classes of hierarchical structures.

The other three papers discuss the large scale inorganic hierarchy, the domain of self-gravitating bodies. E. R. Harrison and Michele Kaufman consider the problem of origin of the levels of structure that are observed in the universe. Harrison reviews modern approaches to cosmology through gravitational theories treating "smoothed" universes, in which the various cosmic sub-structures are replaced by a hypothetical uniform perfect fluid whose density and motion conform to the averages for observed bodies. The difficulties of recapturing the observed structure from a homogeneous fluid in the time span of the accepted age of the universe are developed. Kaufman reports the explanation that she and Layzer derive for the origin of the various levels of self-gravitating bodies based on the role of electrostatic forces in a cold, compact, primordial universe. While their model is successful in producing a sequence of gravitating bodies, it runs into difficulty with the observation of the 3°K background radiation. Albert Wilson views the cosmic hierarchy as a structure he calls a modular hierarchy, i.e., a set of levels each characterized by an aggregate or module that is in turn decomposable into sub-modules associated with the next lower level and grouped into super-modules associated with the next higher level. He shows that among cosmic bodies, the modular levels may be characterized by a density parameter that appears to assume only discrete values. For gravitating systems, density parametization is equivalent to a time parametization, implying that each modular level may be associated with a discrete characteristic time.

In the two notes that follow, Robert Williams illustrates special cases of Euler's Law by aggregating geometrical polyhedra and demonstrates how dimension $(n + 1)$ emerges from the operation of combining entities of n dimension. Paul Shlichta illustrates the existence of overlap in three examples from inorganic hierarchies – symmetry groups, polyhedra and crystal structures – and raises the question of using "tree-like" diagrams to study hierarchical structures.

■

Structural Hierarchy in Inorganic Systems

Cyril Stanley Smith*

Although inorganic materials lack some of the richness of structure of biological and social aggregates, they are nevertheless complex enough to illustrate most of the principles of aggregation.

A NOTE ON HISTORY

Historically, realistic concepts of structure were slow to develop despite an early appreciation of the texture of matter in artisan's fracture tests and the emphasis on the particulate nature of matter by the early atomic philosophers. It seems to have been hard to see crystallinity in anything but externally perfect polyhedra. The immense strides of mathematical physics in the 17th and 18th centuries were possible only by ignoring those properties of matter that depended upon structure. Not until into the twentieth century did the real structure of matter become of concern to physicists, and even molecular chemistry in the 19th century was more a system of notation than it was a statement of real spatial structure.[1]

One of the major achievements of 19th century science was the development of the laws describing the formation of simple molecules, an immediate outcome of Dalton's quantitative atomism beautifully confirmed by volumetric measurements on combining gases and by the superb success of the kinetic theory in explaining their PVT relationships. It was so successful, in fact, that it blinded both physicists and chemists to the importance of crystalline aggregation. One of the most surprising results of the applications of X-ray diffraction to the study of crystals in 1912 was the discovery that most inorganic solids contain no molecules, or rather that

*Department of Humanities, Massachusetts Institute of Technology, Cambridge, Massachusetts, 02139.

1. For an historical analysis see Smith (1960, 1965, 1968). Crosland (1962) provides a fine picture of the development of structural views of the chemical molecule. For a discussion of the role of structure in metallurgy, with photomicrographs of many materials, see Brick, Gordon and Phillips (1965). For an attempt to list and classify all possible microstructures, see Smith (1964). For the structure of crystals on an atomic scale see Bragg (1933), Wells (1956), Evans (1966) and that deservedly popular text by Barret and Massalski (1966).

the whole crystal is the molecule. In sodium chloride, for example, there is no specific pairing of sodium and chlorine atoms any more than there is in solution: The atomic ratios called for by that masterly summary of Daltonian principles, the law of simple combining proportions, are in most cases due to the exigencies of charge and the geometric limitations imposed by the crystal lattice, not to the formation of molecules. The concept of the molecule, once the possibility of changes of its internal structure (allotropy) had been seen, proved flexible enough to account in the minds of 19th century physicists and chemists for all changes of properties. Moreover, the entire realm of organic chemistry and the richness of life itself depends on the almost limitless possibilities of internal arrangement in large molecules that are basically one-dimensional and free from the requirements of overall symmetry. Such molecules can be made to crystallize but they are poor solids, weak and easily melted to give liquids in which molecular identity persists. Irregularly arranged cross linkages between long polymer molecules gave rise to modern plastics, and the field of mixed crystalline and molecular solids is just opening up.

CRYSTALS

A crystal is formed by the extended repetititon of a locally symmetrical association of atoms. Even when only a single kind of atom is involved, some diversity of arrangement is possible. When more than one species is present, both the possible structures and the mechanisms of growth and change become more numerous. Much of the diversity of properties of real materials lies in the variation with composition of the geometric details of aggregation on an atomistic level within the crystal lattice, but this will not be discussed here because it represents only the first step toward hierarchy. Though the same basic symmetries may appear in very different substances, the specificity or tolerance of local arrangements depends much on the nature of bonding. Solids of the greatest hardness are formed when, as in diamond and other covalent crystals, the rigid internal geometry causing the primary interaction between neighboring atoms can provide the symmetry of the whole crystal. Metallic bonding is less strong, but, being less specific, tolerates the substitution of atoms by others in the formation of an alloy and even extensive plastic deformation without destruction of the crystal lattice. Ionic bonding – balanced repulsion and attraction between charged ions – forms the crystals of most inorganic compounds, which cover a range of hardnesses.[2] The geometry of packing depends rather simply on the relative sizes and charges of the ions. Finally, soft, easily-melted organic solids are examples of molecular crystals in which stable molecules are held rather weakly together by Van der Waals' forces without electron exchange.

Crystals, of course, form only at relatively low temperatures. Some short-range structure persists in the liquid state, and even gases commonly consist of molecules (tightly bound groups of a few atoms) because the partial association of their electron clouds results in a lower energy than fully dissociated ones. At higher temperatures these molecules dissociate; at still higher energies the electrons can be successively stripped from the atomic nuclei, and eventually the nuclei themselves can be disrupted.

Rapid cooling from either the state of vapor or liquid sometimes produces a solid that is essentially amorphous in structure. It is almost impossible to produce amorphous structures from monatomic elements: the energy of misfit is large for such purely atomistic disorder and very little rearrangement is needed to produce microcrystalline packing. However, if pre-aggregation into molecules has occurred (as in sulphur and selenium) or if two or more atomic species differing in size or charge are present, disordered interconnections of locally tightly bound units can be accommodated at various angles and distances without much change of energy, and metastable amorphous structures can be produced at low temperature. This, of course, is the basis of the glassy state which can be homogeneous on all scales above that of the atom. Most materials which seem homogeneous on a large scale are,

2. This is not the place to discuss the electronic quantum-mechanical basis for the different types of molecular and crystalline aggregation. See, for example, Kittel (1967) and Pauling (1960).

however, composed of aggregates of small crystals, more or less randomly oriented. This is by far the commonest form of solid matter.

Figure 1 is a picture from an early textbook of physical metallurgy suggesting how crystals grow from independent nuclei by successive addition of block units and eventually impinge upon each other with the formation of irregular boundaries. Figure 2 is a model popular today in which atoms are simulated by tiny uniform soap bubbles (each approximately 300 microns in diameter), drawn together by the surface tension of the liquid on which they float, to form a two-dimensional raft. Most of the bubbles are close-packed in symmetrical environments forming extensive areas of unchanged orientation, but the whole field is not uniform and localized conflict occurs. The energetics (in the model as with real atoms) are such that the misfit is concentrated in lines of high disorder rather than as a uniform diffuse strain. Two types of boundary will be noted in Figure 2 – the highly disorganized form when the angular change is great, with missing and added atoms and atoms of abnormal coordination, and the small-angle boundary consisting of a line of rather widely spaced dislocations (isolated places of abnormal coordination at the termination of rows of atoms). Imperfections consisting of missing "atoms" and foreign "atoms" can also be seen. Imperfections of any kind when randomly distributed in positions where they ignore or cancel each others' fields do not affect the orientation or

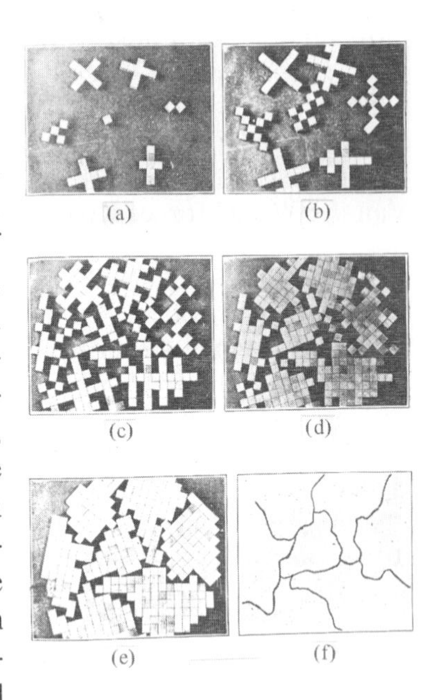

Figure 1. Formation of Intergranular Boundaries (From Walter Rosenhain, Introduction to Physical Metallurgy, London, 1914)

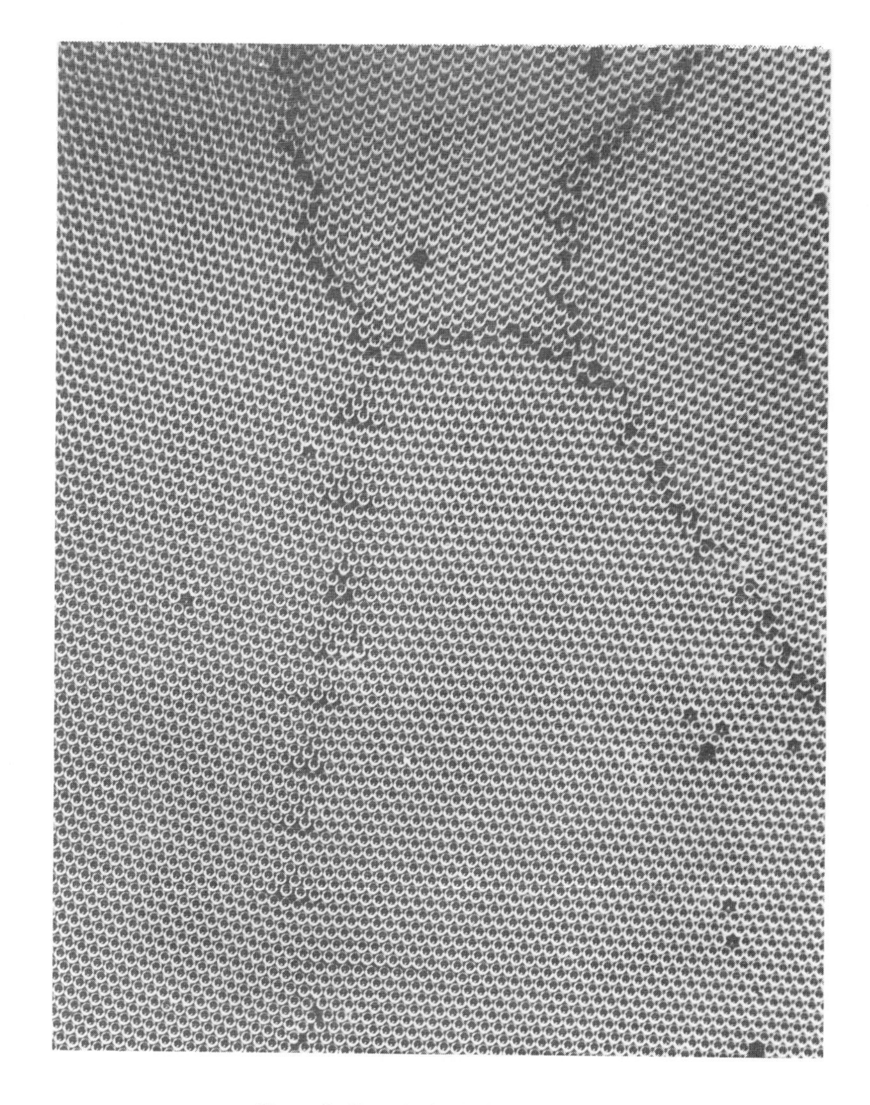

Figure 2. Boundaries in Soap Bubbles

identity of the crystals in which they occur, but if they interact in additively connected fashion a boundary results which separates clearly defined areas.

At lower magnifications, where crystals can be seen but not the atoms, a typical polycrystalline pure material appears in section as in Figure 3. This is a metal, but structures indistinguishable from this occur in ceramics, as well as in sulphides, silicates, carbonates and in many other types of natural rocks. Notice how far removed from the popular image of a crystal such a structure is. The outlines of the crystal grains are more reminiscent of the froth on a glass of beer than of geometric polyhedra: The structure is, in fact, determined by the thin disordered layer of material at the boundaries, which, in most cases, acts like a liquid and has no preferences as to orientation. The boundaries are two-dimensionally continuous throughout the whole aggregate and, to minimize their area, they meet each other as soap films do, in groups of three at dihedral angles of 120° and at the vertices where four grains meet, forming edges making equal angles of 109°.48 with each other in a tetrahedral arrangement. The balance between such angular requirements at the junctions and the necessity for overall space-filling inevitably introduces curvature in the

Figure 3. Microstructure of Pure Aluminum

surfaces. In a soap froth these curvatures are physically balanced by pressure differences, and bubbles with less than 6 sides in a two-dimensional froth (Figure 4) or with less than 14 faces in a three-dimensional froth are under higher than average pressure (Smith 1952).

Here it should be noted that a single individual, whether an atom, a crystal or anything else, can take whatever shape is natural to it; but in an aggregate, the individuals each share an interface with another one in a zone of mutual conformity. Moreover, unless some action occurs across this interface which results in an internal change of structure in both of the contacting units there is no basis for structural aggregation. Although atoms may appear as "hard and shiny balls" when viewed at a large enough scale, they attract each other only because they possess an internal structure that is directly influenced by their external environment. This interaction

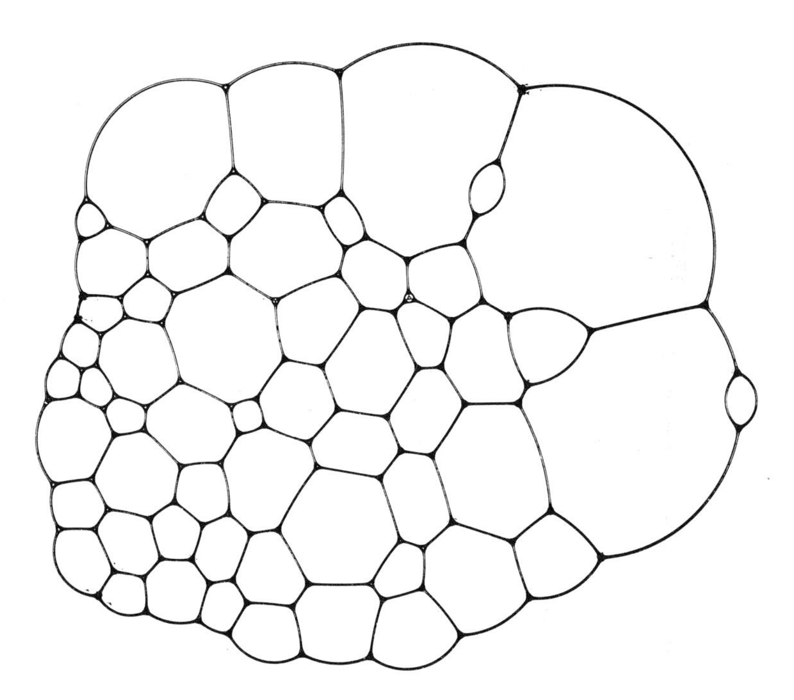

Figure 4. Two-Dimensional Random Froth of Soap Bubbles

between inside and outside as seen by an individual is indeed the very basis of a hierarchy of any kind whatever.[3]

No grain boundary is recognizable in groups of a few atoms for it is a property only of a large enough aggregate for locally divergent group properties to have emerged. Any sufficiently large ordered array inevitably (as a consequence of entropy or, the same thing, as a consequence of the randomness of the history that gave rise to it) has areas of local departure from order. Note that the two-dimensional boundary in a three-dimensional aggregate has come about because of imperfections which are three-dimensional on the atomic level. There is an intimate connection between dimension (number, rather) and dimensionality. Note also that at a high scale of resolution the boundary is fuzzy, indeed its existence becomes a question of definition involving connections beyond the normal field of view; a lower effective magnification will for a time reveal sharp boundaries, but at still lower resolution they again disappear. The existence of any kind of domain is intimately related to the scale and definition of the boundaries used to define them. The converse of this is that an aggregate which on a large scale appears homogeneous with a certain average composition, charge, density or any other property can always be subdivided into space-filling cells, each of which contains a combination of positive and negative deviations to yield a cell of exactly the average property. The connected surfaces which define these "cells" may coincide with physical discontinuities, but are sometimes merely surfaces at which the gradient of some significant property changes sign. It seems impossible to escape from the limit of human perception, in fact, one could claim that hierarchy in the super-atomic world is mainly in our perception; we understand large things mainly by "explaining" them in terms of small ones. If the steps are too small, we lose the advantage of separation, if they are too far apart they

3. This atomistic view of the world, that nevertheless requires internal structure and a means of communication within the "atom" at each level, may not be infinitely extendable; but it does seem to hold at all levels accessible to man. In any case, despite the philosophers, science does not really depend on the elaboration of ultimate principles, but rather starts somewhere in the middle and spreads out with ever-widening interacting concepts which become increasingly probable as they interlock in the well explored regions but are never anchored in fundamental certainity. See the discussion of the inductive maze and the justification of probable inference in Hawkins (1964: pp 289-244; 246-9).

encompass too much complexity to be grasped, but this does not mean that nature is so limited.

STRUCTURAL CHANGE

In systems of any kind, change depends on the formation and/or movement of interfaces. A polycrystalline aggregate can arise in a thermodynamically unstable environment from either the discontinuous nucleation of new crystals and the enlargement of these until they impinge upon each other, or from the gradual condensation of imperfections to leave more perfect regions. In the former case well-defined boundaries move and merge; the system is visibly heterogeneous. In the latter, fuzzily defined regions become progressively sharper. Thermodynamically, the first is a first order, usually high energy, change. The second, of relatively low energy, is second order, with continuous change of structure and energy. The resulting structures may be indistinguishable from each other, for in both cases the process ends by the boundaries adjusting to their own needs. In a pure metal the first corresponds to the growth of crystals from the liquid or from another (unstable) crystalline form, the latter to the recovery of a cold worked material, in which imperfections (dislocations and vacancies) which are at first widely scattered throughout the volume of the strained crystal collect together to form a network of sub-grain boundaries. The sub-boundaries themselves may further condense, but often they are swept away by the movement through them of major boundaries (recrystallization) – a process that on a larger scale appears as a nucleation and growth process. In most cases, when the grains have reached composition equilibrium throughout their volumes, further geometric change occurs simply by movements of the grain boundaries to decrease their area and hence the energy of the system.

The two kinds of change are seen more clearly in alloys containing more than one species of atom, for chemical diffusion must occur to adjust regions to new equilibrium conditions. New crystals form by the nucleation and movement of an interface that sharply separates volumes of different composition; growth is limited by chemical diffusion. Alternately in phase separation by spinodal decomposition – a mechanism recognized by Gibbs in 1878 but which, in solids, has only recently begun to receive the attention that it deserves (Cahn 1968) – there is a gradual sharpening of composition fluctuations in those thermodynamic regions where uphill diffusion is possible, accompanied by the gradual locking-in of positive and negative areas to produce a mutually satisfactory geometry in which each phase is continuous and initially of the same shape as the other. In most cases, because of differences of atom sizes, the structural framework will change as the gradient steepens at first producing balanced regions of positive and negative strain and eventually giving rise to structural interfaces as dislocations are generated. This whole sequence is well shown in the atomic rearrangements accompanying the hardening of duralumin. In art perhaps it is analogous to what has been called organic sculpture in contrast to the more sharply defined growth forms of the constructivists.

A word should also be said about the disintegration of structures, the formation not of an interface across which communication between different areas is maintained, but of a crack, an intrusion of indifferent space between two now-isolated surfaces. This is exemplified by the cracking of a mudflat, the shattering of a highly strained brittle material or the building of an automobile throughway across a natural landscape or a settled community. Such cracks are rarely formed by condensation; usually they form by branching growth of a dominating intrusion under conditions that preclude volume adjustment.

CRYSTAL LATTICES

A full treatment of inorganic hierarchy would include much about the internal symmetries of crystals, and the mathematical basis of the 230 classes of point group symmetry. The determination of the precise location of atoms within the crystallographic unit cell, possible only after the discovery of X-ray diffraction, is one of the central themes of twentieth century solid-state physics. These structures result from a balance between the angles and distances demanded by the interactions of the outer electrons of neighboring atoms with the space-filling requirements of an extended array of identical groupings. Local and long range symmetry interlock. Local clusters of atoms may have any shape, but if a crystal is to be formed, small groups of such clusters must pack to form a unit cell. To fill space when stacked together in extended symmetry the unit cells must either be cubes, prisms, or one of the rectilinear distortions of these (topologically unchanged) which compose the fourteen Bravais lattices.

The geometric relations are highly important, but it is topology rather than geometry that determines the hierarchical possiblities. In the past, symmetry has properly been emphasized for reasons of both experimental and mathematical simplicity – and perhaps for aesthetic reasons also – but it tends to disguise the physical nature of aggregation. The atoms see their neighbors, not the symmetry rules of a crystal. Not only is it a manifestly unphysical representation of the sharing of bonds between neighboring atoms to cut them into unit cells as in the frequent representation (Figure 5); but emphasis on the unit cell, essential of course for the building of the crystal, disguises the fact that in most classes of substance the basic physical group is either smaller or larger, that is, a molecule, a domain, or an entire crystal. In solid solutions local symmetries within the cells differ enormously.

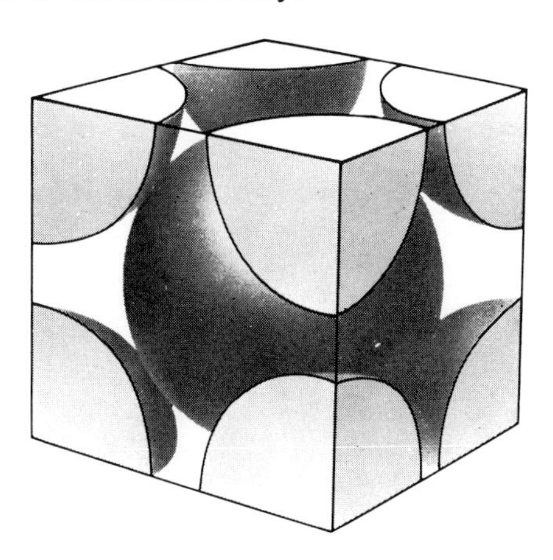

Figure 5. Atom Packing in a Body-Centered Cubic Crystal

DOMAINS AND INTERFACES: THE STRUCTURAL BASIS OF HIERARCHY

In a crystalline material composed of only one type of atom (i.e., an element), the only steps to a larger structure are the breaking up of the lattice by an orientation change as shown in Figures 6 and 7 (analogous to Figure 2), or by the formation of a boundary between different coexisting allotropic modifications, shown in Figure 8a in the simplest form between two areas differing only in spacing. Figure 8b emphasizes the coordination change at the interface.

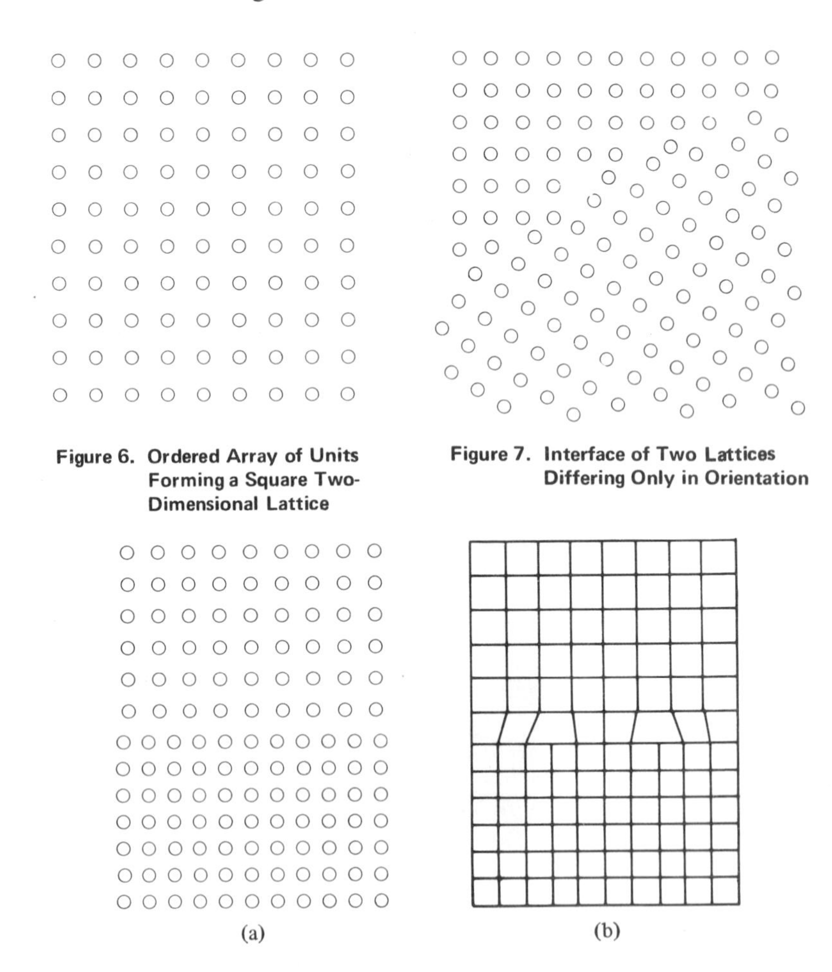

Figure 6. Ordered Array of Units Forming a Square Two-Dimensional Lattice

Figure 7. Interface of Two Lattices Differing Only in Orientation

Figure 8. Junction of Two Lattices Differing Only in Spacing

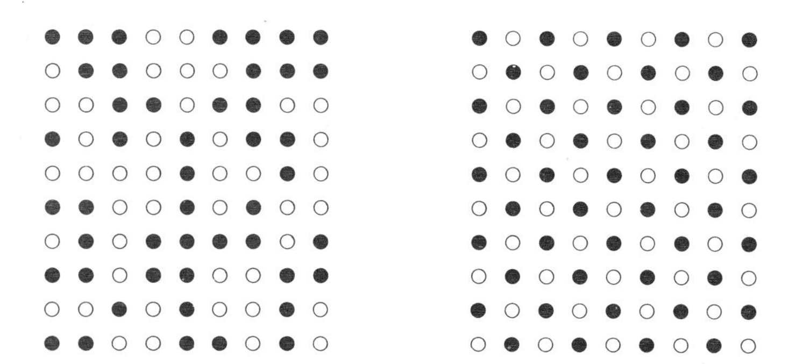

Figure 9. Equal Number of Black and White Atoms Randomly Distributed

Figure 10. Equal Number of Black and White Atoms in an Ordered Arrangement

When two or more atom species are present, hierarchical structures can develop even within the lattice of a single crystal. Figure 9 shows a simple square lattice with equal numbers of black and white "atoms" arranged randomly, as occurs physically when the difference of energy between various associated pairs is overcome by the randomizing influence of thermal agitation. On any scale above the atom such an array will appear completely homogeneous – no line can be drawn that is distinguishable from a parallel one and there is no basis for separating one zone from another.

Figure 10 shows a lattice with the same "atoms" as Figure 9 in an ordered arrangement such as exists in a sodium chloride crystal with positive and negative ions arranged in the lowest energy configuration. Here alternate atom planes in some directions are distinguishable, but on any larger scale this too is homogenous. However, even such simple order gives rise to the possibility of another kind of discontinuity to separate and identify regions of structure on a larger scale. Thus if, keeping order and orientation constant, one zone is displaced an odd number of atom steps from its neighbor, a domain boundary consisting of neighboring pairs of atoms in abnormal coordination results from the change of step (Figure 11). Note, however, that no boundary will form if the out-of-step regions are local and randomly distributed (as in Figure 9) so that they

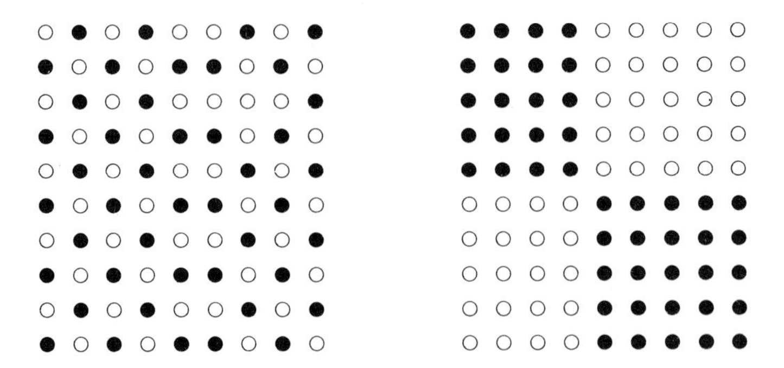

Figure 11. Out-of-Step Domain Between Two Regions

Figure 12. Compositional-Segregation of Equal Number of Black and White Atoms

do not interact to form a connected boundary. Three-dimensional space cannot be divided into cells unless there is some two-dimensional feature marking a discontinuity on the scale of observation.

Figure 12 illustrates the structure when the "atoms" have segregated into separate areas uniformly of one kind. In this case unlike all the earlier ones there are differences in gross chemical composition as well as in arrangement in the local regions. Such structures are heterogeneous on some scale considerably above that of the atom—typically, in metallurgical and mineralogical cases, on the scale corresponding to the resolution of the optical microscopy, though these too seem homogeneous at the scale of the naked eye.

The origin of the distinction between the physical structures exemplified by Figures 9 to 12 lies in the relative energies of interaction between like and unlike atoms. Figure 9 corresponds to the case where bond energies are indifferent, Figures 10 and 11 to the case where bonds between dissimilar atoms are stronger than between similar ones (but note that an out-of-step domain becomes increasingly costly as the inequality becomes greater), and Figure 12 to the case in which like neighbors are preferred. Usually, of course, segregation such as in Figure 12 also involves a change of structure, and the interface involves a geometric and/or topological discontinuity superimposed in the chemical one. With more than two types of atoms and in structures of lower symmetry, more complicated subgroupings are possible but in all cases (unless the distinguishing characteristic is randomly distributed) there will be at some scale regions of homogeneity separated by an interface.

A characteristic of the structures considered (indeed a characteristic of *all* structures with an emerging structure at a higher level) is that the boundaries can be revealed by translation. If the structure is transposed a unit distance and redeposited upon itself, a moiré pattern will appear at the boundaries regardless of whether the boundary marks changes of order, orientation, composition or any other detectable characteristic, while similar displacements within each zone will cause no change. This is perhaps the best test for the existence of any interface and for hierarchical possibilities. (Translation parallel to the interface will not reveal it, hence test translations must be done at three orthogonal directions in a three-dimensional system.)

In the above discussion the black and white spots in Figures 9 to 12 were assumed to be chemically different atoms. In this sense Figure 9 illustrates solid solution, Figure 10, an ordered alloy or ionic crystal, Figure 11, out-of-step domains in an ordered alloy, and Figure 12, segregation. However, the same relationships apply to groups of any units that are distinguishable in any way. Thus, the black and white spots can equally well represent electron spin direction (up or down) in magnetic materials – in which case Figure 9 corresponds to a paramagnetic material, Figure 10 represents antiferromagnetism, Figure 11, an antiferromagnetic domain wall and Figure 12, ferromagnetic domains. If the symmetry relations are partly or completely relaxed to give a network more like a soap froth than a crystal lattice similar diagrams can also represent the structures at lower magnifications, the black and white spots then representing whole domains or crystal grains of differing orientations and/or of differing compositions.

In a less well defined sense, diagrams of this kind can represent the structure of almost anything that exists, even patterns of perception, knowledge and thought. For in every case there are definable and repeatable local interactions, connections and avoidances which are repeated to some extent, but among which discontinuities eventually emerge and give rise to larger groupings by their interrelations. Even groupings of individuals in human society follow similar structural principles, for if there are more than two they interact positively or negatively with each other on the basis of sex, interlocking skills or needs, religious or political opinions, national or cultural inbreeding, or a host of special interests. The analogy is hard to see because of the overlapping of many different structural elements and interactions in a given individual, and the fact that the change of structure is usually in idea space not in the movement of physically identifiable bodies. But, in any one dimension, the principles of local interaction leading to conflict with other locally homogeneous groups is clear enough, as is the parallel with reconstructive (revolutionary) heterogeneous nucleation and the less disruptive gradual change by cumulative change of opinion as in spinodal phase separation.

SOME ELEMENTARY TOPOLOGY

The great mathematician Leonhard Euler in 1752 saw that the number of faces, edges and vertices on a polyhedron were simply related. His law was reformulated in more general terms by Schläfli in 1852. For our purposes it can be given as follows:

$$N_0 - N_1 + N_2 - \ldots + (-1)^{n-1} N_{n-1} = 1 - (-1)^n, \quad (1)$$

where N_0, N_1, N_2, N_3 represent the number of zero, one, two and three-dimensional features in a simply connected array.[4] This relationship results directly from the sharing of vertices (N_0 or C) and edges (N_1 or E) in the connected network, and is clearly basic to any aggregate. A rudimentary example appears in Figure 13. In an *extended* simply-connected array of polygons there is an inescapable relationship between the numbers of lines meeting at each vertex (r) and the number of

4. An excellent treatise on the mathematics of aggregates is given by Coxeter (1948 and 1961).

sides (n) to the polygon, and, the number of faces (N_2 or P) in three-dimensional stacks of polyhedra. For example, Figure 4, represents a two-dimensional array with three and only three films meeting at every junction.[5] Except for edges (E_b) on the outer boundary of the array, each polygon (P) with n sides contributes $n/2$ edges to the array. Since three edges (E) meet each vertex (C), $3C = 2E$ and equation (1) simplifies to:

$$\Sigma (6 - n)P_n - E_b = 6. \qquad (2)$$

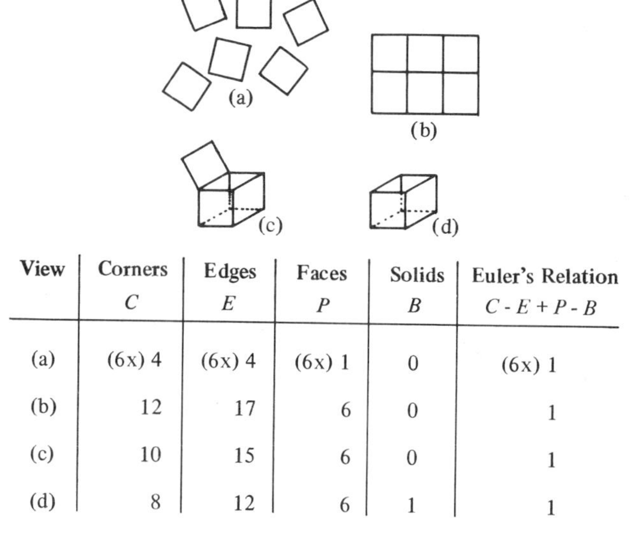

View	Corners C	Edges E	Faces P	Solids B	Euler's Relation $C - E + P - B$
(a)	(6x) 4	(6x) 4	(6x) 1	0	(6x) 1
(b)	12	17	6	0	1
(c)	10	15	6	0	1
(d)	8	12	6	1	1

Figure 13. A Rudimentary Explanation of Euler's Law

5. In three-dimensions if the minimum number of junctions is specified, as in a soap froth, all corners are tetrahedral, i.e., there are 4 three-dimensional, 6 two-dimensional and 4 one-dimensional features meeting at each vertex. If the angles at which the edges meet are to be equal they must be 109°.48 (the angle whose cosine is minus one-third) which does not correspond to any plane polygon. Descartes' rule applied to the vertex angle requirements shows that the impossible ideal polyhedron would have 5.1043 sides, 13.394 faces and 22.789 corners. As Kelvin showed long ago, the best solution to space filling with a single polyhedron of minimum surface area is a 14-sided cube-octahedron having 6 four-sided faces and 8 six-sided ones, the latter being doubly curved to permit them to join at the correct angle. In a random soap froth, and in many other undifferentiated aggregates such as metal grains and biological cells pentagons are by far the most frequent polygon (Smith 1952).

In any two-dimensional set of polygons such relations are most simply summarized in the following equation:

$$\Sigma\, rC_r = \Sigma\, nP_n + E_b \qquad (3)$$

If polygons are to be capable of filling space by simple repetition, then (correcting fractionally for the sharing of edges and corners with adjacent polygons) the following relation between the average number of sides of polygons and average number meeting at each vertex must hold:

$$\frac{C}{P} = \frac{\bar{n} - 2}{2} = \frac{2}{\bar{r} - 2} \qquad (4)$$

For identical polygons, only triangles, squares, and hexagons can conform to these requirements, and these are the only bases of an extended identical array. However, assorted polygons of any type can almost always be grouped into larger tessellations that are either hexagonal or quadrilateral and these will be conformally duplicable *ad infinitum.* (This is the topological equivalent on the crystallographer's unit cell.) Some simple symmetrical groups can be seen on the Chinese lattice designs of Figure 14, but even completely random arrangements such as Figure 15 can be subdivided into either hexagonal or quadrilateral groupings (often both) composed of different assortments, subject always to the requirement that the above equation be satisfied within each. The hierarchy of such groupings of polygons on larger and larger scales parallels both the emergence of physical discontinuity as in Figures 6 to 12 and the eventual averaging out of structure and properties. It merits detailed mathematical analysis, especially in three dimensions.[6]

6. The following relations, derived from Euler's Law, apply to all two-dimensional simply-connected nets of polygons cut from an infinitely extendable net. With appropriate corrections for sharing at the severed edge, they will be useful in pinpointing departures from a normal coordination of nearest neighbors, for they give zero weight to certain specific features. For example, equation (a) shows the relations between the numbers of various nonhexagons in a soap bubble array, for not only is $\Sigma\,(6-2r)$ then zero but hexagons do not count; equation (d) relates nonrectangles with departures from four-fold coordination, equation (b) gives the number of polygons solely as a function of the number of corners of different types. The change of such characteristics which occur when groupings of polygons are

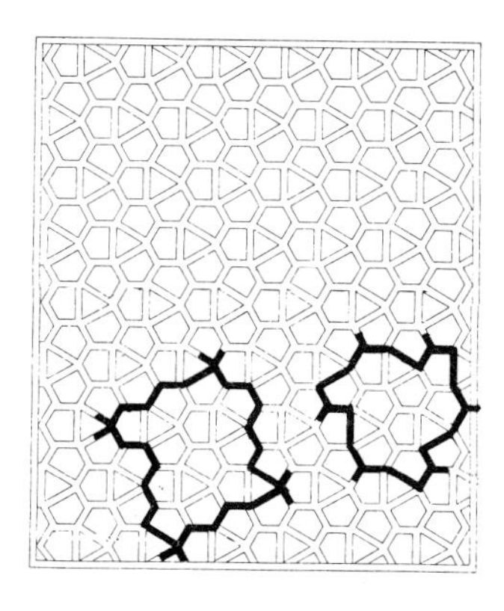

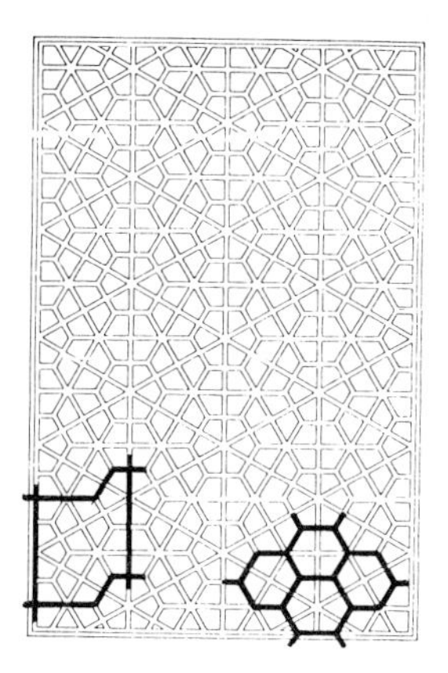

Figure 14. Symmetrical Groupings Superimposed on Chinese Lattice Designs

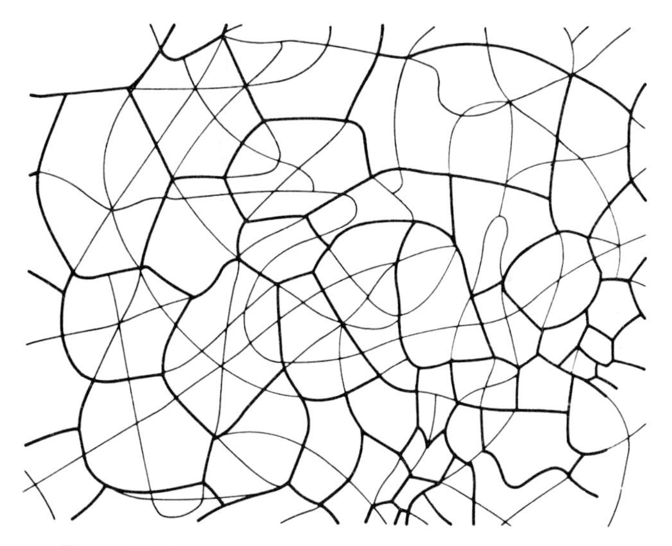

Figure 15. Hexagonal Net Superimposed on Random Array

enlarged and with the displacement of test boundaries is the mathematical equivalent of the moiré pattern effect mentioned above.

(a) $\Sigma\, rC_r = \Sigma\, nP_n$

(b) $P = \Sigma \left(\frac{r}{2} - 1\right) C_r$

(c) $\Sigma\,(n-2)P_n = 2C$

(d) $\Sigma\,(n-4)P_n = \Sigma\,(4-r)C_r$

(e) $\Sigma\,(n-6)P_n = \Sigma\,(6-r)C_r$

SOME PRINCIPLES OF HIERARCHY

Consideration of such structures as the above have led the writer to attempt the enunciation of some hierarchical principles. Though composed with simple inorganic aggregates in mind, I believe them to be rather generally applicable.[7]

In the following paragraphs the unit in general is any three-dimensional element that is identifiable and is capable of interaction with others approximately its own scale. The terminology is that of the one, two and three dimensions of matter in ordinary space, though most of the statement may be more broadly interpreted, for it is part of the definition of structure that any structure, whether material, biological, social or intellectual must be composed of some identifiable parts having some relationship to each other.

The world is complex and understanding is simple; however, there is always some scale on which significant interaction involves relatively few units. Complexity comes from the factorial combinations of these units. The new structures that seem to emerge as aggregates when seen on a larger scale are partly illusory: it is less a characteristic of the structure itself than of the limited resolution of our perception (whether visual or conceptual). Each "level" is what we see at certain resolutions, and corresponds to the matching of only those structural elements that can be resolved without too much

7. Victor Weisskopf in his delightful *Knowledge and Wonder* (New York, 1962) describes the "quantum ladder," with the energy associated with a unit of change within a structure becoming larger as the scale becomes smaller. At any energy density there is some clumped structure that acts as if it were indestructible. Normally solid crystals and stable organic molecules are easily melted or dissociated with everyday energies, but our world would not exist were not the chemical atom, or its nucleus rather, effectively immutable. It should be noted, however, that the small energies associated with structure changes in large aggregates are large changes for those few atoms that are principally affected. Thermodynamic averaging disguises the historically-determined real differences between complex structures. At the other extreme scientists' attention has focused on simple associations involving such large energy changes for each unit that historical diversity is also erased. The ladder of structural hierarcy is the inverse of the quantum ladder, but above the atom the levels are less well defined, and therefore more subject to perceptual resolution and decision than those of higher quantum energies. It is usually possible to follow relationships both up and down the ladder for two or three rungs from any point of origin. Some degree of arbitrariness is usually involved: it is the old philosophical issue as to how many grains of sand make a heap. Nature itself comprises all levels and knows not our distinctions.

detail at a single effect viewing distance. Both the narrow cone of sharp vision and the simplicity of the fact or idea-patterns that we can have in our minds at any one time introduce a basic indeterminacy in our knowledge of the world. Yet we can sequentially apply this attention span to many different things on one scale. We can use the microscope to diminish, or distance to increase the effective scale and to some extent, through a combination of memory and forgetfulness, or a wilful lack of precision, we can relate all these views and find larger patterns. That patterns perpetually reappear at each stage of resolution is a simple result of the fact that at each scale, the most energetic as well as the most visible interactions are between few neighbors and there are not many significantly different arrangements possible between them, however diverse the whole assembly may be. In the past, science has tended to ignore the downward interplay between scales and concentrated analytically on simple selected details at a single scale; in the future it will be necessary to develop hierarchical methods for studying the two-way interaction between levels.

Some general principles seem to be the following:

1) An assembly of units will not form a coherent aggregate unless the parts interact in such a way as to modify their internal structure and energy. An organism is an aggregate in which there is a dominant center of organization and significant internal communication. (A physically unconnected assembly can be related intellectually as a structure in the mind of an observer.)

2) Any aggregate that is neither completely ordered nor completely disordered must have hierarchical aspects, but the perception of the levels of the hierarchy requires the recognition of a two-dimensional surface to define each three-dimensional unit in accordance with Euler's Law. In general, the surface is defined by interactions between discontinuities in the coordination of the parts on the next lower level; discontinuities not so connectable constitute merely internal features. The number of imperfections, and hence the number of possible interfaces, increases with size for simple reasons of entropy.

3) Each scale of structure has a type and energy of interaction appropriate to it. Structure and energy are inseparable. Though the strongest interactions between any units are those between neighbors on the same scale, residual effects extend downwards through a few and upwards through many levels—but not indefinitely, for beyond a certain point complexity becomes irresolvable and merges again with homogeneity. A human being is not far from the point of maximum significant complexity that can be usefully associated with quantum interaction between chemical atoms, while the planets and their satellites are not far from the smallest gravitationally-determined structures.

4) Complex structures cannot originate instantaneously but are formed in time: they must have had a history. A complex structure is both a partial record of past history and a framework within which future changes occur by the operation of physical laws.

5) Structural change occurs by the formation of interfaces and/or by their movement. New interfaces can form internally by the progressive condensation of imperfections causing the self-enhancement of fluctuations until the gradients are sharp enough to become definable interfaces. The external growth of an individual unit occurs by accretion which causes the translation of an existing interface. The latter is typical of crystal growth, the former of spinodal transformation, biological cell division, convection cells and stream-bed formation.

6) A complex aggregate may contain superimposed several different structural hierarchies (each corresponding to a different type of interaction between units, based on a different aspects of internal structure). The hierarchies may be partially independent or their various levels may interfere constructively or destructively with each other. In inorganic aggregates all interactions, not necessarily those between adjacent atoms, are those of quantized photons transmitting electrical or magnetic fields. In biological organisms there are superimposed more complex interactions based on mass transfer of chemical messengers which react only with specific complex local

arrangements; in intellectual or social systems small and large thought patterns interlock and new patterns arise in the constructive mismatching of communicated parts of the old.

CONCLUSION

One of the most significant aspects of the above discussion is the realization that connected disorder on one level of organization gives rise to significant pattern on a higher level. Historically it was the artist who discovered the symmetries that arise in the repetition of identical motifs regardless of their shape, and it was he who exploited the importance of variability within a framework of order. He has a greater sense of the relationship of parts to wholes than do most scientists. Many artists have exploited the fact that there are basically very few unit patterns involved in any combination. Practically everything can be divided into isolated granular areas with concomitant connectivity of the boundaries, and branching connectivity of linear growing tree-like forms. For instance, a Chinese landscape painter of the 12th century was likely to use the building up of brush strokes to define the crystalline texture of a rock in the foreground, while similar shapes and textures reappear in relation to boulders in the middle ground and to mountains in the distance. Forking cracks in the rocks are related to trees, to branching rivers, to mountain valleys and to the gathering or inverse branching of mountain ridges. The local relationships are simple, but multiplied in combination they diversify almost without limit, only to repeat at each scale something of the pattern that existed before. Perhaps the scientists' concern with accuracy of statement, which by its very nature must be limited, needs to be tempered a little more than it has been in recent years with the artist's awareness of larger scale interactions.

ACKNOWLEDGEMENTS

The author's thoughts on the nature of complex structures have been deeply influenced by conversations stretching over many years with

Professors Victor Weisskopf and John Cahn. He is indebted to the latter also for an unusually constructive criticism of a draft of the present paper, and to Michael Bever for helpful comments.

REFERENCES

Barrett, C. S. and Massalski, T. B. 1966. *The Structure of Metals.* 3rd ed. New York: McGraw-Hill.

Bragg, W. L. 1933. *The Crystalline State: Vol. 1, A General Survey.* London: G. Bell.

Brick, R. M.; Gordon, R. B.; and Phillips, A. 1955. *Structure and Properties of alloys.* 3rd ed. New York: McGraw-Hill.

Cahn, J. W. 1968. "Spinodal Decomposition." *Trans. Met. Soc. AIME* 242:168-80.

Crosland, Maurice. 1962. *Historical Studies in the Language of Chemistry.* Cambridge: Harvard University Press.

Coxeter, H. S. M. 1948. *Regular Polytopes.* New York: MacMillan.

——. 1961. *Introduction to Geometry.* New York: John Wiley & Sons.

Evans, R. C. 1966. *An Introduction to Crystal Chemistry.* Cambridge: Cambridge University Press.

Hawkins, David. 1964. *The Language of Nature: An Essay in the Philosophy of Science.* San Francisco: Freeman & Co.

Kittel, C. 1967. *Introduction to Solid State Physics.* 3rd ed. New York: John Wiley & Sons.

Pauling, Linus. 1960. *The Nature of the Chemical Bond.* 3rd ed. Ithaca, N. Y.: Cornell University Press.

Smith, C. S. 1952. "Grain Shapes and Other Metallurgical Applications of Topology." In *Metal Interfaces* pp. 65-108. Cleveland: Am. Soc. Metallurgy.

——. 1960. *The History of Metallography.* Chicago: University of Chicago Press.

Smith, C. S. 1964. "Some Elementary Principles of Polycrystalline Microstructure." *Metallurgical Reviews.* 9:1-48.

——. 1965. "The Prehistory of Solid State Physics." *Physics Today* 18:18-30.

——. 1968. "Matter versus Materials: A Historical View." *Science* 162:637-44.

Wells, A. F. 1956. *The Third Dimension in Chemistry.* Oxford: Clarendon Press.

■

The Mystery of Structure in the Universe

E. R. Harrison*

WHY THE MYSTERY?

A universe without planets, stars, and galaxies seems inconceivable. Almost all the structure of the physical world that we take for granted would cease to exist. Yet strangely enough cosmology has not yet succeeded in explaining in general terms why the universe is fragmented into planets, stars, and galaxies.

Hopefully, the formation of planets will be better understood when we know more about star formation. Star formation studies draw upon the full richness of the interior environment of galaxies, and in spite of many unsolved problems, there is a feeling of moderate confidence among astrophysicists that they are proceeding along the right lines. It is believed that the origin of planets and stars can in principle be explained, and that there are no insurmountable problems in the way. This whole approach, of course, takes for granted the existence of galaxies with their complex interiors of enhanced and irregular density that is so necessary for forming stars.

Without galaxies to set the stage it is unlikely that there would be any stars or planets. Thus the secret of most kinds of astrophysical structure is concealed ultimately in the problem of understanding the origin and formation of galaxies. But so far all efforts to explain the origin of galaxies have failed; we do not know how galaxies have evolved nor what initial conditions determined their present properties and structural differences.

Various problems of galaxy formation have in recent years become issues of major importance in cosmology. (Harrison 1967b and Field 1969). What were the initial conditions in the remote past that launched the expanding universe on a career of fragmentation? What determines the masses of galaxies and

---

**Department of Physics and Astronomy, University of Massachusetts, Amherst, Massachusetts 01002.*

their morphological differences? What creates the initial conditions? Because we do not know the answer to these and various other questions, many of which pose problems that seem nowadays insurmountable, it is true to say that the cause of all large scale structures in the universe is veiled in mystery. To explain why galaxy formation is so mysterious it is necessary to say a little about cosmology.

THE OLDEST OF ALL SCIENCES

The oldest of all the physical sciences is almost certainly cosmology. It is the science of the universe as an organized whole. Nobody in their right mind for a moment dreams that we are anywhere near achieving even an elementary notion of the universe as an organized whole. Because its aim and scope are so sweeping and breathtaking, cosmology remains among the most backward sciences in spite of its antiquity. At present it is the Cinderella of the sciences; we may hope that one day it will ultimately take its rightful place as queen of the physical sciences.

Because the universe is so complex we must commence by simplifying it and discarding what hopefully we believe to be the irrelevant aspects. We start by throwing out planetary systems, and in an all or nothing spirit follow by dismissing the entire range of stellar and galactic structure. Everything is smeared into a uniform fluid, and we are left with an idealized universe that is virtually little more than the grin on the face of a Cheshire cat. Just before all structure is dissolved away we manage to seize hold of some of the grosser rudiments of the cosmos. The universe tends to be isotropic about us, both in distribution of mass and in recession, and what we know of it furthermore does not contradict the possibility that it is homogeneous. The featureless universe has thus the property of being invariant under rotation and translation in space. These properties or symmetry postulates have far-reaching consequences. For example, they restore the notion of absolute motion and also speeds (of recession) can once again exceed the speed of light. It is not surprising that in cosmology general relativity yields results that are sometimes indistinguishable from those of Newtonian theory.

A uniform cloud of gas expanding under the influence of gravity obeys the equation

$$\frac{1}{2} v^2 = \frac{GM}{r} - C, \tag{1}$$

where v is the velocity at a given radius r, $M = 4\pi\rho r^3/3$ in the mass interior to radius r, and C is a constant of the motion. If the cloud is to remain uniform in density at all instants, then all distances in the cloud must increase by the same amount. Thus if r_0 is a radius at some instant, then at all other instants the radius is

$$r = r_o \frac{R}{R_o}$$

when R varies with time, but is the same everywhere in space. The velocity at radius r is therefore

$$v = \frac{r_o}{R_o} \frac{dR}{dt},$$

and equation (1) can be written

$$\left(\frac{dR}{dt}\right)^2 = \frac{8\pi G\rho R^3}{3} - \kappa \tag{2}$$

where $\kappa = 2CR_0^2/r_0^2$ is a new constant. This constant can be adjusted by altering the value of r_0 and we choose r_0 so that κ is either 0, +1, or −1. These are the famous Friedmann models for which the variation of R with time is shown schematically in Figure 1. With equation (2) we must also have an equation of state that shows how pressure varies with energy density. Using the symmetry postulates discussed above and elementary thermodynamics it is found that

$$\frac{d}{dt}\left(\rho R^3\right) + \frac{p}{c^2} \frac{dR^3}{dt} = 0. \tag{3}$$

This is equivalent to $dU + pdV = 0$, and $U = \rho c^2 V$ is the internal energy of an element of expanding volume of arbitrary size. Equations (2) and (3) are the equations of a uniform universe as derived by general relativity. As we see, they can also be derived by simpler arguments (Bondi 1960; Irvine 1965; Callan et al., 1965; Harrison 1965).

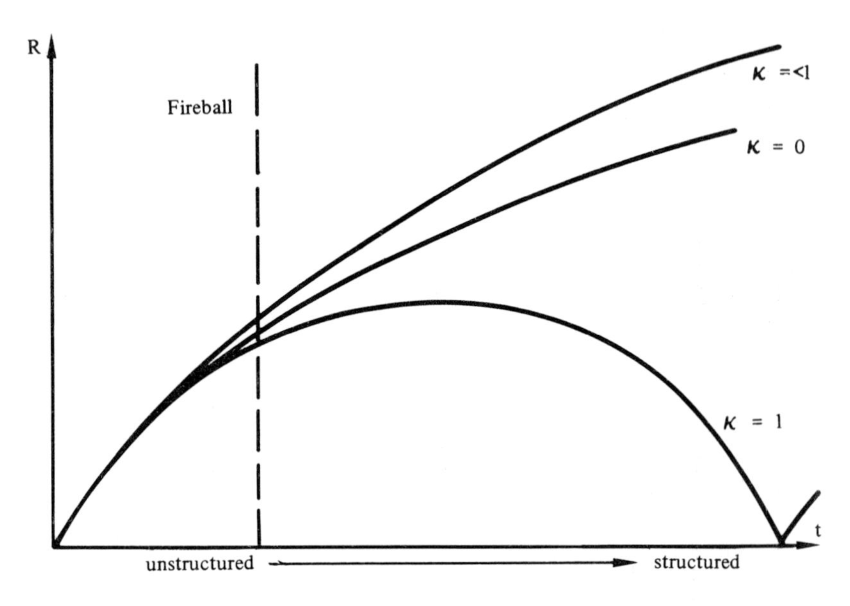

Figure 1 Friedmann Models

THE BIG BANG

We shall not discuss the differences between the three possible models of an idealized universe. Cosmology for many years has devoted itself to discussions of this type, and to inspired guesses as to which of the three simplifications is the most appropriate representation of the physical universe. It is noticed, however, that as we go back in time towards the origin at $t = 0$, all three models converge and become in many ways similar to each other. The density of the universe rises and eventually approaches infinity as t goes to zero. What happens as we approach this singularity no one knows; classical cosmology apparently breaks down and the physics of the universe verge on chaos in the midst of incredible complexity (Wheeler 1957; Harrison 1967a).

The Friedmann type models all involve an initial phase of high density and temperature, and are referred to as the big bang models. The initial phase, or fireball, terminates when the universe is approximately 10^6 years old and matter consisting mainly of protons and electrons begins to emerge as the dominant constituent of the universe (Dicke, et al., 1965; Harrison 1968b). The continual expansion of the universe during 10^{10} years has cooled the immense energy of the fireball to the present 3° K black body radiation.

It seems unbelievable that structure of any kind can exist within the explosive fury of the fireball. One would expect, as we look back into the past and see the density and temperature rise to trillions of times greater than in any star, that all variation of any kind is crushed into a uniform whole. This is the general view: the early unstructured universe expands and evolves into its present highly structured state. It is also a comforting view in the sense that structure becomes an evolutionary detail, and is not a basic issue in cosmology, thus justifying to some extent the use of idealized and featureless models.

THE PROBLEM OF PUTTING STRUCTURE BACK INTO THE UNIVERSE

We have created smooth and featureless models of the universe consistent with the view that initially the universe is unstructured. The next step is to show that irregularities grow and in the course of time the unstructured universe becomes the structured universe, as indicated in Figure 1.

I should hasten to say at this point that in this discussion our main concern is with macroscopic physics. In everyday life we take for granted the fact that the physical world consists of complex arrangements of matter. The diversity and richness of the planetary environment is explained by the physics of small scale and large scale structures: of short range relatively strong interactions on the atomic and nuclear scale, and of the long range relatively weak force of gravity that becomes dominant on the stellar and galactic scale. The problem of incorporating microstructure in the universe is immensely difficult. Eddington (1946) labored boldly to show that elementary particles were of direct cosmological significance. Nowadays we know more about the properties of elementary particles and consequently are less bold. Many physicists believe even so that cosmology and high energy physics are converging onto areas of common interest wherein the very small and the very large have much in common.

In the meanwhile a more modest program seeks to account in a natural way for the presence of macrostructure on the galactic scale. We must in effect recover in an evolutionary manner the structure which earlier was so lightly discarded in the idealized models. Unfortunately, all attempts to restore structure to the universe have so far failed. The impression grows that we have discarded too much, and have paid too high a price for the idealized models; not only has the bath water been thrown away, but the baby also. The expanding structureless universe is barren and tends to remain uniformly bleak. In the following section we consider ways and means of injecting structure into the universe, and show that the mystery is not so much in the mechanisms of growth as in the initial conditions that are postulated.

COSMOGONIES ANCIENT AND MODERN

The Jeans' Cosmogony. It is thought by many that the problem of fragmenting the expanding universe into protogalaxies is physically similar to the problem of star formation in galaxies. Small density disturbances under certain conditions tend to grow. Many authors have derived criteria of instability, and have proceeded immediately to confer structure on the universe in the arm-waving manner of a conjurer.

Imagine a spherical surface embedded in the cosmic fluid and that its radius increases as the universe expands. The Jeans' criterion tells us that if the velocity of the expanding surface is less than the velocity of sound, then the matter inside the sphere is stable against small density perturbations. However, if the surface velocity is more than the velocity of sound, then the enclosed matter is unstable. In this case small random fluctuations in density will grow, that is $\delta\rho/\rho$ will grow, but not necessarily the fluctuation $\delta\rho$. The rate of growth of such density disturbances in the expanding universe turns out to be very slow, and consequently it is often necessary to postulate a foundation of initial irregularity as large as $\delta\rho/\rho = 10^{-2}$. The dissipation of this type of irregularity in the fireball is an additional weakness of the theory. The Jeans' cosmogony offers us an amplification mechanism requiring however an input of previous structure at reduced amplitude. The objection to this type of cosmogony is the tendency to claim that all is solved. It is implied that the utter simplicity of the proposed initial irregularity in some way mitigates its large nonthermal magnitude. The universe, however, is not always quite so simple as our theoretical cosmic fluid, and it is quite likely that the initial conditions are far more complex than is allowed for within the limitations of the hydrodynamic prescription. By sacrificing the simplicity of the Jeans' cosmogony it is possible to devise initial conditions of smaller amplitude. Their complexity might hopefully shed light on such perplexing problems as galactic rotation and galactic magnetic fields.

Lagging and Delayed Cores. A promising cosmogony is the following (Bonnor 1956; Ne'eman 1965; Novikov 1965; Ne'eman and Tauber 1967). We imagine that a region of the cosmic fluid expands more slowly than the rest of the universe. This is called a lagging core, as shown in Figure 2. An alternative argument assumes that the region lies dormant as a singularity, and commences to expand later than the rest of the universe at epoch t_0. This is called a delayed core, and is also shown in Figure 2.

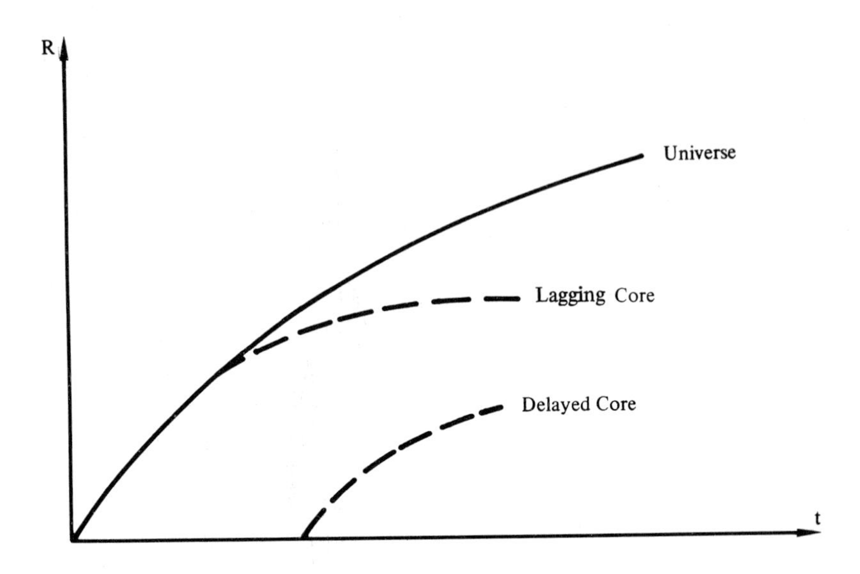

Figure 2 Two Kinds of Kinematic Inhomogenity

The initial conditions of these cores are evidently not just simple density fluctuations of the cosmic fluid. More appropriately one might refer to them as kinematic irregularities – certain regions possess less energy and therefore expand more slowly. The great advantage of this type of inhomogeneity is that its growth rate is moderately rapid and it develops into advanced structural forms at an early era. By taking into account time retardation, due to gravitational redshift, it is possible for lagging cores to evolve within the fireball and retain the properties of the fireball long after the radiation dominated era has finished (Harrison in preparation). Lagging cores of this kind possess remarkable properties; they consist mainly of radiation and resemble Wheeler's thermal geons (1962). As their mass diminishes because of radiation losses, the redshift also diminishes, and the combination of these processes means that the luminousity slowly rises and then reaches rapidly a peak value in the late universe. In effect, the lagging core has become a little bang.

The lagging core concept is still in an early stage of development. A great attraction of the theory is that it provides a plausable explanation of those extraordinary astronomical events in which large amounts of energy are released catastrophically, as in exploding galactic nuclei and quasars. Alternative explanations so far have stressed the idea of gravitational collapse and the release of nuclear and gravitational energy in supermassive objects, or have elaborated on such themes as collisions and supernova explosions in compact stellar systems. Lagging cores have the advantage of possessing the immense energy density of the fireball and there is no difficulty in meeting the energy requirements of quasars.

Matter and Antimatter. In the very early universe the temperature is high enough to create pairs of all the particles with which we are familiar. This means that the universe then consisted mainly of matter and antimatter. Their difference was extremely small: matter exceeded antimatter by only one part in a billion (Harrison 1968a). We have seen that one of the outstanding unresolved problems in cosmology is the origin of structure. A second outstanding problem is the reason why matter should be slightly favored compared with antimatter.

Both these problems can be resolved if we adopt initial conditions different from those previously discussed. Instead of assuming density or kinematic variations, we turn to the possibility of composition variations. In the early universe the only composition variation of any consequence is the difference between matter and antimatter. Thus in some places matter slightly exceeds antimatter, and in other places antimatter slightly exceeds matter. Thus the problem of their difference in amount is resolved by proposing that it is an inhomogeneity that exists on the local scale but not the cosmic scale.

As the universe expands the temperature gets lower and pair annihilation progressively eliminates matter and antimatter. Eventually, all that remains is their difference and a large amount of radiation. But the difference in some places is matter, and other places antimatter. Thus as the universe expands and cools it also fragments into islands which are the protogalaxies. The composition inhomogeneity has the virtue that it kills two birds with one stone. Furthermore, it can be shown that composition variations lead in a natural way to kinematic variations and all the phenomena associated with lagging cores.

IN THE BEGINNING

In cosmology we start by shutting our eyes to the array of astrophysical structures in order to make some progress. But the progress has been so slow and the returns for our sweeping idealizations so little, that we are forced to wonder whether the slate has not been wiped too clean. This view is reinforced by the discovery that structure cannot reappear spontaneously in the featureless models. Structure evolves out of primoidal initial conditions that exist as inherent inhomogeneities in the universe from its beginning, or at least from the first moment of its expansion. Moreover, the initial conditions are a foundation of structure that in all probability possesses some degree of complexity as a natural and essential part of the universe.

The symmetries of cosmology, such as the translational and rotational invariance in space, are in fact too simple. The existence of primitive structure from the beginning means that these symmetries are never perfect; their violation is presumably allowed and made essential within the scope of higher and more general symmetries which so far have not been discovered.

We can summarize by saying that in the last few years the study of galaxy formation has made a certain amount of progress, if only in the direction of discovering the immense complexity that surrounds the subject. Without knowing how galaxies originate, our cosmologies are in nothing less than a pitiable state. And there is also an important gap in our understanding of all astrophysical structures which are related to or derive from the existence of galaxies. Our discussion has tended to emphasize our ignorance and to show that the mystery which surrounds the problem is associated with the puzzling nature of the initial conditions. Out of this approach there emerges, however, an exciting prospect. Structure on the large scale does not develop in the post-fireball era from random fluctuations and suddenly spring to adulthood; but rather it develops in the early universe from moderately complex rudiments of deep cosmological significance, and already in the fireball era there exist well-developed configurations whose variety and complexity foretell of the intricate universe that lies billions of years ahead in our own time.

ACKNOWLEDGEMENT

This research was supported in part by the National Science Foundation. Also published as Contribution Number 24, 12 November 1968 of the Four-College Observatories.

REFERENCES

Bondi, H. 1960. *Cosmology.* Cambridge: Cambridge University Press.

Bonnor, W. B. 1956. "Formation of the Nebulae." *Zeitschrift für Astrophysik* 39:143-59.

Callan, C.; Dicke, R. H.; and Peebles, P. J. E. 1956. "Cosmology and Newtonian Mechanics." *Amer. J. Phys.* 33:105-8.

Dicke, R. H.; Peebles, P. J. E.; Roll, P. G.; and Wilkinson, D. T. 1965. "Cosmic Black-Body Radiation." *Astrophys. J.* 142:414-19.

Eddington, A. S. 1946. *Fundamental Theory.* Cambridge: Cambridge University Press.

Field, G. B. 1969. Chapter 12 in *Galaxies and the Universe, Vol. IX, Stars and Stellar Systems,* eds. A. and M. Sandage. Chicago: University of Chicago Press.

Harrison, E. R. 1965. "Cosmology without General Relativity." *Ann. Phys.* 35:437-46.

——. 1967a. "Quantum Cosmology." *Nature* 215:151-52.

——. 1967b. "Normal Modes of Vibrations of the Universe." *Rev. Mod. Phys.* 39:862-82.

——. 1968a. "Baryon Inhomogeneity in the Early Universe." *Phys. Rev.* 167:1170-75.

——. 1968b. "The Early Universe." *Physics Today* 21:31-39 (June).

Irvine, W. M. 1965. "Local Irregularities in an Expanding Universe." *Ann. Phys.* 32:322-47.

Ne'eman, Y. 1965. "Expansion as an Energy Source in Quasi–Stellar Radio Sources." *Astrophys. J.* 141:1303–5.

Ne'eman, Y. and Tauber, G. 1967. "The Lagging-Core Model for Quasi–Stellar Sources." *Astrophys. J* 150:755-66.

Novikov, I. D. 1965. "Delayed Explosion of a Part of the Friedmann Universe, and Quasars." *Soviet Astronomy–AJ* 8:857–63.

Wheeler, J. H. 1957. "On the Nature of Quantum Geometrodynamics." *Ann. Phys.* 2:604–14.

——. 1962. *Geometrodynamics.* New York: Academic Press.

■

A Possible Mechanism for the Origin of the Sequence of Cosmic Bodies

Michele Kaufman*

INTRODUCTION

Whether or not astronomical objects form a hierarchy in the rigorous sense of the definition given by Bunge (1969) may be open to question. I shall not deal with this topic except to point out that the classes of astronomical bodies form a sequence of objects of increasing mass. Usually, the members of any one class in the sequence move under the gravitational influence of higher members of the sequence – for example, planets move in orbits around stars. Planets also move with stars around the center of the galaxy. Members of any one class may also be components of the members of higher classes in the sequence, but not necessarily the only components. Galaxies are composed of stars, but galaxies also contain uncondensed gas, dust, cosmic rays, and radiation. It is also possible, though not as yet observed, that stars may be found outside of galaxies in intergalactic space.

There are various ways in which the classes of cosmic bodies can be compared. As one goes from stars to galaxies, the average density decreases. We could also compare the ratio of the mass to the radiation emitted. In this paper, however, I discuss a possible mechanism for the formation of self-gravitating bodies rather than compare their properties.

The sequence of astronomical bodies terminates with clusters or possibly superclusters of galaxies. The next largest system appears to be the universe itself. If we average the density over suitably large volumes, the distribution of matter in the universe seems to be homogeneous and isotropic (that is, independent of direction). On the other hand, we know that the distribution of matter in the universe at the present time is nonuniform on a local scale. The average density of matter within a star is much greater than the average density in interstellar space; the average density within a galaxy is much greater than the average density of matter in the space between

---

*Department of Physics, Brown University, Providence, Rhode Island 02912.

galaxies. Was the distribution of matter in the universe always nonuniform in this sense? Or did the universe start out homogeneous at high densities, and galaxies develop from what originally were chance concentrations of particles as the universe expanded?

To postulate that the universe began with significant nonuniformities which developed into galaxies as the universe expanded is to assume rather special initial conditions. Theorists, therefore, have devoted considerable effort to determine whether galaxies can form from random density fluctuations in an initially homogeneous universe. According to standard theories of the expanding universe, it takes 10^{10} years for the universe to expand from a density where atomic nuclei are in contact to its present density. Therefore, any theory of galaxy formation which is consistent with standard evolutionary cosmologies must be able to produce galaxies within 10^{10} years.

Another observational feature that must be accounted for is shown in Figure 1 which is based on data in Allen (1963). Not

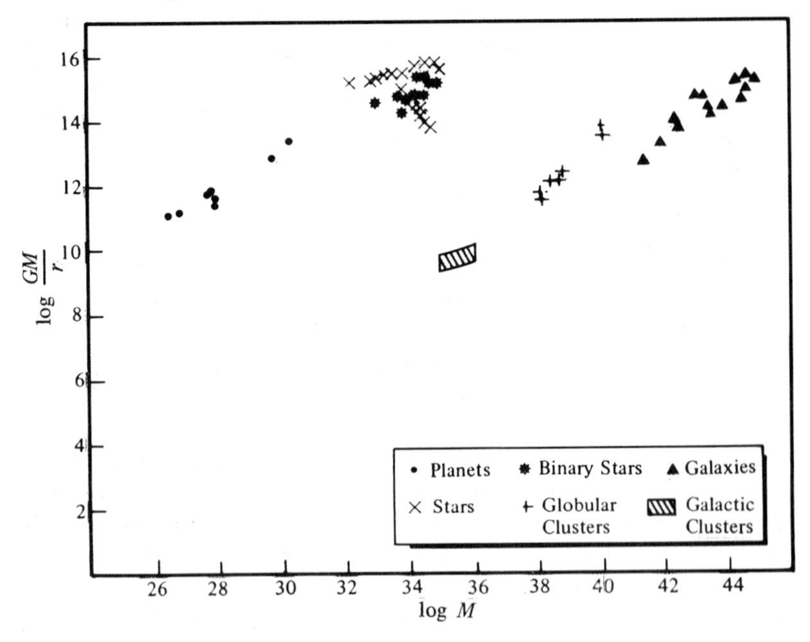

Figure 1 Gravitational Energy per unit Mass as a Function of Mass

only is the distribution of matter locally nonuniform, but we observe a wide variety of self-gravitating bodies ranging in scale from planets to galaxies that possess gravitational binding energies per unit mass within a fairly narrow range of values. The binding energy of a stable bound configuration of particles is the minimum energy required to disperse the components of the system to infinity. This might suggest that similar physical considerations would account for the existence of all members of the astronomical hierarchy. Whether or not this is the case, the fundamental problem is to find a mechanism which would explain how at least the large scale self-gravitating bodies (i.e., galaxies and clusters of galaxies) could have separated out of the expanding universe and attained binding energies per unit mass of the order of 10^{13} to 10^{15} ergs/gram.

Is gravitational attraction between particles sufficient to bring about the formation of galaxies? According to calculations by Lifshitz (1946), Bonnor (1957), Layzer (1964), Peebles (1965), Hawking (1966), Eltgroth (1966), Harrison (1967) and others, gravitational forces are too weak by many orders of magnitude to cause random density fluctuations in an initially homogeneous universe to grow into galaxies in a Hubble expansion time of 10^{10} years. A much, much longer time is required. Thus, to account for the existence of galaxies, it appears that one needs either

1) nonuniform initial conditions, or

2) some nongravitational mechanism, or

3) a cosmology which would allow a much longer time scale for the production of self-gravitating systems.

We concentrate here on a nongravitational mechanism for initiating the formation of self-gravitating bodies. This mechanism applies to the formation of any self-gravitating body in the astronomical sequence, regardless of its mass. To be specific, we circumvent the slow initial growth of inhomogeneities under gravitational forces by allowing for electrostatic interactions. The significance of electrostatic forces was suggested Layzer (1964, 1967a, 1967b, 1968b), and the calculations discussed below were performed in collaboration with him. Obviously, electrostatic forces by themselves cannot solve the whole galaxy-formation problem, because in a plasma which is macroscopically neutral, electrostatic forces are effectively short-ranged. However, since the electrostatic attraction between a proton and an electron is approximately 3×10^{39} times stronger than the gravitational attraction between them, electrostatic forces may be sufficiently capable of magnifying statistical irregularities in the density field to initiate the formation of condensations. In fact, we find that under certain specified conditions, electrostatic interactions can lead to instabilities during the early stages of an initially uniform universe.

The physical picture is one in which local concentrations of matter start by expanding slightly less rapidly than the universe as a whole until eventually they become sufficiently differentiated from the general background to evolve independently.

BASIC ASSUMPTIONS AND EQUATIONS

I shall discuss a model for the universe which satisfies the following general assumptions.

1) Einstein's theory of General Relativity applies.

2) The universe satisfies the cosmological principle, i.e., apart from local irregularities, the universe is homogeneous and isotropic.

3) The universe has expanded from densities greater than or equal to nuclear density.

These assumptions allow us to describe the large scale dynamics of the universe. We treat the universe as a completely homogeneous and isotropic fluid called the substratum. Local condensations are considered to be a first order perturbation (Irvine 1965) and do not significantly affect the rate of expansion. Friedmann's solutions of Einstein's equations (Tolman 1934) apply to the substratum. We define a cosmic time, t, as the proper time measured by clocks which move with the substratum. When the pressure, P, is small compared to the rest-mass energy and the cosmic time is also small, then the scale factor, S, (the ratio of a distance at time, t, to a distance at an initial time, t_0) varies as $t^{2/3}$. This relation is exact for all values of t considered here if the curvature is zero.

In addition to the above assumptions, I choose a model which satisfies the following initial conditions assumed to apply at nuclear density.

1) Density is uniform with only statistical fluctuations.

2) Thermodynamic equilibrium prevails. That is, the rate of expansion is so slow compared to collision and reaction rates, that there is sufficient time for the universe to approach thermal and chemical equilibrium.

3) No primordial magnetic fields.

4) No significant amounts of anti-matter.

In an expanding gas, we can distinguish between the energy of the expansion and the random energies that individual particles possess. Let E represent the total energy density measured relative to the expanding substratum. E does not include the expansion energy. In a bound system, E is locally negative, while in a uniform universe, E is everywhere positive. Contributions to E include random kinetic energies, electrostatic and gravitational potential energies associated with local irregularities, radiative energy, and internal "chemical" energy (such as ionization energy):

$$E = E^{kin} + E^{el} + E^{grav} + E^{rad} + E^{int} . \qquad (1)$$

Similarly, the pressure can be written as

$$P = P^{kin} + P^{el} + P^{grav} + P^{rad} , \qquad (2)$$

with

$$P^{rad}/E^{rad} = P^{el}/E^{el} = P^{grav}/E^{grav} = 1/3. \qquad (3)$$

One can show that the potential energy terms do not begin to make important contributions to the energy density until electrons become nonrelativistic. At higher densities, the universe simply cools adiabatically except for contributions from elementary-particle and nuclear reactions. Therefore, let us consider densities at which electrons are nonrelativistic. Then

$$P^{kin}/E^{kin} = 2/3. \quad (4)$$

The pressure and the energy density are also related by a local first law of thermodynamics

$$\frac{d}{dt}(ES^3) + P\frac{dS^3}{dt} = 0, \quad (5)$$

which can be derived from the field equations of general relativity. The inclusion of gravitational contributions to the energy density in this equation has been justified by Irvine (1961, 1965), Layzer (1963), and Dmitriev and Zel'dovich (1964), while the extension to Coulomb forces is given in Kaufman (1968). We can give explicit expressions for the various energy contributions that appear in equation (1).

To simplify the rest of the discussion, I assume a pure hydrogen gas. Calculations by Kaufman (1968) have shown that even if the universe began at zero temperature, nuclear reactions and the decay of hyperons would probably liberate enough energy to make nucleons nondegenerate. (The size of this contribution to the energy depends on the composition of the universe at super-nuclear densities and on the amount of pre-stellar helium produced, if any.) However, if the universe were initially cold enough for matter to be completely degenerate, the energy liberated would be insufficient to lift electron degeneracy. Consequently, when electrons are nonrelativistic let us assume that we can treat nucleons as nondegenerate, but still allow for the possibility of electron degeneracy.

COULOMB INSTABILITY

Under the above conditions, Layzer (1967b) has shown that one can neglect gravitational terms, because Coulomb interactions are much more important than gravitational interactions when only statistical fluctuations are present. Also, initially, the gas is essentially fully ionized by pressure ionization since the density is high. A gas may be treated as fully ionized provided the computed value of the continuum depression due to plasma effects exceeds the ionization potential.

The behavior of the solution to equation (5) depends on how hot the universe is; that is, it depends on the value of T/T_{deg}, the ratio of the temperature to the electron degeneracy temperature, when electrons first become nonrelativistic. If the early universe is hot enough to be radiation-dominated and it cools adiabatically with expansion, the ratio of the electrostatic energy density to the thermal energy density is independent of cosmic time until the temperature drops so low that recombination sets in. Thereafter, the ratio of electrostatic to kinetic energy decreases as the gas recombines. Thus a radiation-dominated universe cools too slowly for electrostatic interactions ever to become important for galaxy formation.

On the other hand, if the universe is sufficiently cold, then Coulomb interactions can produce interesting effects. Figure 2 exhibits the behavior of the kinetic and electrostatic energies per unit mass as functions of cosmic time if the initial value of the ratio T/T_{deg} equals 0.6 at an electron density of $10^{26}\,cm^{-3}$. As the universe expands, the kinetic contribution to the energy density E diminishes relative to the Coulomb contribution. In the typical low temperature case shown, the gas remains pressure-ionized until after the Coulomb energy curve crosses the thermal energy curve. In Figure 3, E_m, the total energy per unit mass relative to the substratum, defined by the relation

$$E_m = E/\bar{\rho}, \tag{6}$$

is plotted versus cosmic time for initial values of T/T_{deg} equal to 0.6 and 2.0. Notice that if one chooses the lower of these two temperatures, then E_m becomes negative, reaching a value of nearly -10^{12} ergs/gm, whereas if one starts with the higher temperature, then recombination sets in while E_m is still

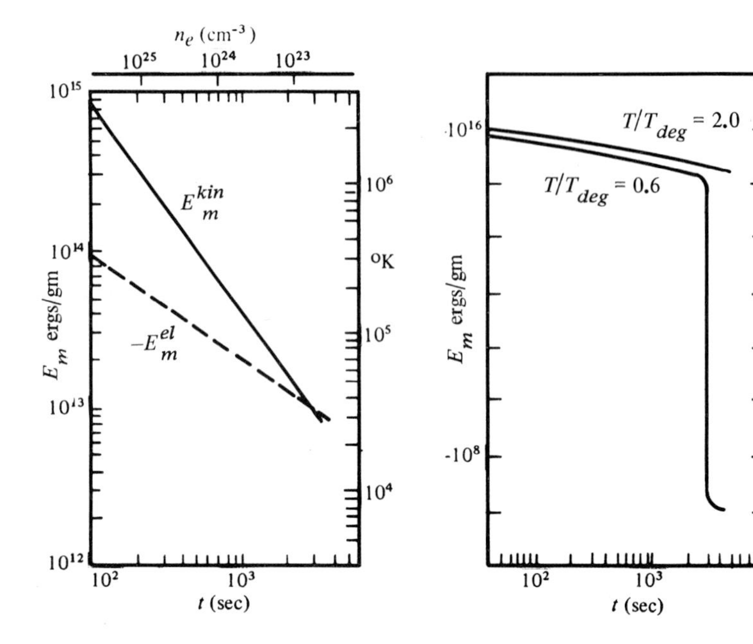

Figure 2 Time Variation of Kinetic and Electric Energies per unit Mass

Figure 3 Energy per unit Mass as a Function of Cosmic Time

positive. One can show that if the universe is cool enough so that the ratio $T/T_{deg} \lesssim 0.7$ when electrons become nonrelativistic, then Coulomb interactions can drive the total energy density E negative.

Condensations do not form as soon as the total energy becomes negative, because the isothermal compressibility is still finite. However, if, at the epoch considered, the universe is so cold that the value of the ratio T/T_{deg} is less than 0.3, then the cosmic gas can reach a critical point in the pressure-density phase diagram. At a critical point, the isothermal compressibility becomes infinite. Provided the temperature is nonzero, the Ornstein-Zernicke theory (Landau and Lifshitz 1958) predicts that at the critical point, macroscopic density fluctuations of large amplitude can occur on all scales. In the present context, only irregularities larger in scale than the Jeans' wavelength but smaller than the radius of the observable universe develop appreciably. Equation (5) governs the amplitude and growth rate of density fluctuations.

Layzer (1968b) argues that once the universe has reached a critical point, the tendency of the cosmic distribution of matter to relax to a quasi-stationary state will be opposed by the formation of macroscopic density fluctuations which tend to lift the instability. As a result, the pressure P and the energy density E remain close to their critical values until self-gravitating systems begin to separate out. As long as E^{rad} is small and the gas is fully ionized, the solution for the behavior of the gas as gravitational energy is released at the critical point is straightforward. The results are:

1) the gas heats up and becomes nondegenerate in a time scale of the order of the Hubble expansion time at that epoch;

2) E_m is of the order of -10^{13} ergs/gm and, as the density decreases, becomes proportional to the scale factor, S;

3) the electrostatic energy decreases in importance as the local gravitational energy increases in importance;

4) the pressure becomes positive;

5) the gravitational potential energy per unit mass associated with local irregularities changes rapidly from a near-zero value appropriate to a uniform gas with random fluctuations, to values of the order of -10^{13} ergs/gm, close to that observed in present-day self-gravitating bodies. Although the change in E^{grav} occurs rapidly, it is not too fast for energy to be supplied by gravitational contraction of objects with radii less than the radius of the observable universe at rates that are small compared to the free-fall rate.

Just prior to the time t^* at which our model reached a critical point, we can calculate E_m^{grav} by taking values for the density contrast (the ratio of the density fluctuation to the density) and the correlation distance appropriate to thermal fluctuations. Almost immediately after critical point instability sets in, the product of the correlation distance and the density contrast, increases by a factor of 10^{16}. If the correlation distance becomes equal to the radius of the observable universe at this epoch, then the density contrast has increased to a value of approximately 10^{-5}, which is a factor of 10^{28} greater than the density contrast for thermal fluctuations of this size. Note that a condensation with this radius would have a mass of approximately 10^9 solar masses, about a factor of 10 less than the mass of our galaxy. This rapid increase in the density contrast just as electrostatic interactions drive the universe to a critical point is many orders of magnitude faster than would be the case if the universe did not reach a critical point. Subsequent increase in the density contrast proceeds much more slowly and becomes proportional to the scale factor, S, as long as the gas remains completely ionized and radiative terms are small. Calculations to describe the behavior of the density contrast when these conditions no longer apply are in progress. When the density contrast approaches unity, it is expected that self-gravitating systems will separate out of the expanding universe.

IDEALIZED MODEL FOR FORMATION OF SELF-GRAVITATING SYSTEMS

Layzer (1968b) has given a more explicit criterion for when self-gravitating bodies form. He has shown that a *necessary* condition for the formation of a self-gravitating system of radius R is

$$U_m^{self}(R) \doteq |E_m^{grav}(R) \tag{7}$$

where $U_m^{self}(R)$ is the gravitational energy per unit mass of a uniform self-gravitating system of radius R. To derive a *sufficient* condition, one must omit from the right hand side of this equation all gravitational energy associated with condensations that exist with masses smaller than $M(R)$. The model discussed has the property that an incipient condensation of given mass may contain condensations of smaller mass and may be a member of an incipient condensation of larger mass. This criterion for the formation of self-gravitating systems may be expressed as an integral equation whose solutions are discrete values. One obtains a sequence of self-gravitating bodies, with systems of larger mass forming at later stages in the expansion.

Layzer (1968b) finds that simple analytic model for the dependence of $E_m^{grav}(r)$ on r gives a sequence of systems in which M_n, the mass of the nth system, is related to M_1, the mass of the first system, by

$$M_n/M_1 = 2^{3n}, \tag{8}$$

and

$$U_m^{self} \propto M^{1/3}. \tag{9}$$

As one can see from the slope equal to 1/3 in the distribution of globular clusters and galaxies shown in Figure 1, the relation between the gravitational potential energy of globular clusters and the gravitational potential energy of galaxies agrees with equation (9).

Thus we have a possible mechanism for the formation of the hierarchical sequence of cosmic bodies in a model of the universe in which the density field is initially uniform. This mechanism works only if the initial universe is sufficiently cold. However, if the universe was not initially very hot, then one cannot attribute the present observed isotropic microwave radiation to a primordial radiation field. Alternative suggestions, which might account for the creation of such a radiation field at more recent epochs have been given by Layzer (1968a) and by Low and Kleinmann (1968).

REFERENCES

Allen, C. W. 1963. *Astrophysical Quantities.* 2nd ed. London: Athlone Press.

Bonnor, W. B. 1957. "Jeans' Formula for Gravitational Instability." *Mo. Notices Roy. Astron. Soc.* 117:104-17.

Bunge, M. 1969. "The Metaphysics, Epistemology and Methodology of Levels" (this volume).

Dmitriev, N. A., and Zel'dovich, Ya. B. 1964. "The Energy of Accidental Motions in an Expanding Universe." *Soviet Physics JETP* 18:793-96.

Eltgroth, P. 1966. "Fluctuations and Correlations in the Expanding Universe." Thesis: Harvard University. Smithsonian Astrophys. Observ. Special Report 197.

Harrison, E. R. 1967. "On the Origin of Structure in Certain Models of the Universe." *Mem. Soc. Roy. Liège* 15:15-28.

Hawking, S. W. 1966. "Perturbations of an Expanding Universe." *Astrophys. J.* 145:544-54.

Irvine, W. M. 1961. "Local Irregularities in a Universe Satisfying the Cosmological Principle." Thesis: Harvard University.

——. 1965. "Local Irregularities in an Expanding Universe." *Ann. Phys.* 32:322-47.

Kaufman, M. 1968. "Thermal History of the Early Stages of a Matter-Dominated Universe." Thesis: Harvard University.

Landau, Ld., and Lifshitz, E. M. 1958. *Statistical Physics.* Reading, Mass.: Addison-Wesley.

Layzer, D. 1963. "A Preface to Cosmogony. I. The Energy Equations and the Virial Theorem for Cosmic Distributions." *Astrophys. J.* 138:175-84.

——. 1964. "The Formation of Stars and Galaxies: Unified Hypothesis." In *Annual Review of Astronomy and Astrophysics*, Vol. 2, eds. Goldberg, Deutsch and Layzer, pp. 341-62. Palo Alto, Calif.: Annual Reviews, Inc.

——. 1967a. "Formation of Astronomical Systems in an Expanding Universe." *Mem. Soc. Roy. Liège* 15:33-38.

——. 1967b. "Relativity Theory and Astrophysics." *Lectures in Applied Mathematics, 8.* American Math Society.

——. 1968a. "Black-Body Radiation in a Cold Universe." *Astrophysical Letters* 1:93-102.

——. 1968b. "Cosmogonic Processes." Lectures at Brandeis Summer Institute.

Lifshitz, E. M. 1946. "On the Gravitational Stability of the Expanding Universe." *J. Phys.* (U.S.S.R.) 10:116-29.

Low, F. J., and Kleinmann, D. E. 1968. "Infrared Observations of Seyfert Galaxies, Quasars, and Planetary Nebulae." *Astronom J.* 73:868-69.

Peebles, P. J. E. 1965. "The Black-Body Rediation Content of the Universe and the Formation of Galaxies." *Astrophys. J.* 142:1317-26.

Tolman, R. C. 1934. *Relativity, Thermodynamics and Cosmology.* Oxford: Clarendon Press.

■

Hierarchical Structure in the Cosmos

Albert Wilson*

The primary focus of cosmological thought in the present century has been on interpreting the observations of the sample of the universe available to our telescopes in terms of a set of models based on various theories of gravitation, especially the General Theory of Relativity. The problem of the structure *of* the universe is customarily divorced from the problem of the structure *in* the universe. Theoretical cosmologists usually choose to explain the structure and behavior – past and future – of the universe with models that smooth out the distribution of matter in the universe, replacing the observed structured distribution of matter with a uniform homogeneous perfect fluid whose density varies in time, but not in space. However, the structure contained *in* the universe becomes difficult to relate to models constructed around smoothing postulates. This has resulted in separate theoretical approaches to the origin of the various structures in the universe. While most of these approaches have met with some success, they are inadequately related to one another and to cosmological theories.

The arbitrary separation of the structure and behavior of the universe from the structure and behavior of its contents may be expedient from the point of view of mathematical simplification, but it cannot be accepted as more than an exploratory strategy. The observational tests for discriminating between various cosmological models are difficult and marginal. Since several smoothed models are candidates for best fit to the observations, it is unfortunate that the large amount of information contained in the sub-structures of the universe cannot be used in testing these models. But until models that relate the properties of the sub-structures to the properties of the whole are employed, much information of potential cosmological value in sub-structure astronomical observations is not cosmologically useful.

**Douglas Advanced Research Laboratories, Huntington Beach, California 92647.*

So long as the cosmological problem has been approached through smoothing out the sub-structures, it is not surprising that little attention has been paid to the regularities that exist among the sub-structures. There are many features of the visible sample of the universe that suggest that the regularities in sub-structures which range over 40 orders of magnitude in size and 80 orders of magnitude in mass, are of central significance to the order and operation of the universe. The fact that these regularities may not be readily explainable in terms of existing physical theories, should not deter their examination. The object of this paper is to present an overview of the known structural regularities that link the properties of physical bodies across a hierarchy of levels from the atomic to the cosmic.

MODULAR HIERARCHIES

Because of the confusion created by the many uses of the term "hierarchy" some amplification concerning the sense in which hierarchy is used in astronomy and cosmology is needed. Astronomical usage, in general, employs "hierarchy" to mean a *set of related levels* where the levels may be distinguished by a size or mass parameter. Examples from the past include the hierarchy of spheres associated in ancient cosmographies with the various heavenly bodies beginning with the moon and continuing to the sphere of fixed stars, and the hierarchy of epicycles used by Ptolemy to account for observed planetary motions. Modern concepts of hierarchy in the cosmos began with the speculations of Lambert (1761) who extrapolated to higher order systems the analogy between a satellite system such as that of Jupiter and its moons and the solar system of the sun and its planets. Lambert speculated on a hierarchy consisting of a distant center about which the sun orbited as a satellite and an even more distant center about which the first center orbited, and on to more and more distant centers comprising larger and larger systems. To explain Olbers' and Seeliger's Paradox; Charlier (1908, 1922) posited a universe built up of a hierarchy of "galaxies." The first order galaxies were the familiar ones composed of stars, second order galaxies were aggregates of first order galaxies, third order of second order, and so on. Shapley (1930) pointed to the set of levels into which all matter appears to be organized extending from the sub-atomic particles to the "metagalaxies." Shapley's organization, like Charlier's, constructed the material bodies on any level from the bodies on the level next below. A hierarchy of this type which is of fundamental importance in astronomy we designate a *modular hierarchy*.

The central idea in a modular hierarchy is the *module* which is a structure or a system that may be regarded both as a *whole*, decomposible into sub-modules identified with a lower level, and as a *part* combinable into super-modules identified with a higher level. In astronomy, even though the modules on any level are not identical, the levels may be readily distinguished on the basis of the nature of the principal sub-modules out of which entities are directly composed. Thus, for organization in a modular hierarchy, open and globular star clusters and galaxies would be assigned the same level, all being aggregates of stars. Stars, planets, and moons, all built from atoms, would share the next lower level, while clusters of galaxies would be assigned the next level above. There are several other ways than that of a modular hierarchy for organizing cosmic bodies into levels. Some of these will be discussed later.

The term "module" being used here in this general sense need not be precisely defined, however, we may ascribe two fundamental properties to modules. First, a module possesses some sort of closure or partial closure (Wilson 1969). This closure may be topological, temporal, or defined by some operational rule as in group theory. Second, modules possess a degree of semi-autonomy with respect to other modules and to their context. These two properties appear to be common in all modular hierarchies.

In considering the origin of a modular hierarchy we may inquire at any level as to whether the size, the complexity, and the limits to the module are determined (1) totally by the properties of its sub-structures, (2) by its environment, or (3) by a combination of both module contents and context. And to these logical possibilities we must add a fourth: that the levels and modules in a hierarchical structure are determined by some principle or process that operates independently of all levels of the hierarchy. In this fourth case the *levels* of the modular hierarchy themselves become the *modules* on a single level of a meta-hierarchy. The various levels in the meta-hierarchy are an observable level, an energy or force level and a meta-relational level. As an example, we may think of the lines in the spectrum of an atom as an ordinary hierarchy (but not a modular hierarchy). The levels of the meta-hierarchy would be the spectral lines, the energy levels, and the mathematical law – such as the Balmer formula – that defines the sequence. It may be objected that this is but a representational hierarchy. But the essential point is that the levels are neither determined by the sub-levels nor the super levels, but by a set of eigen values that act as a causal meta-relation.

COSMIC-ATOMIC NUMERICAL RELATIONS

Let us now return to our specific example of a modular hierarchy: the levels of cosmic structure. Instead of assuming a two level model of the cosmos – the level of a homogeneous perfect fluid and the level of the universe as a whole – we shall attempt a multi-level view retaining the atomic, stellar, galactic, galaxy cluster and universe levels. Further, in view of the lacunae in our knowledge of physical processes governing "vertical" relations between levels, it is appropriate to work from observation toward theory. In doing this the steps we must take are somewhat analogous to those taken by Kepler and his successors in the investigation of planetary orbits. From the arithmetic ratios of various powers of the sizes and periods of planetary orbits, Kepler discovered his kinematical relations and from these later came Newton's formulation of the physical laws governing planetary motions. Thus while our ultimate goal is the formulation of the physical laws and processes governing the relations between the levels in the cosmic hierarchy, our immediate goal is much more modest. It is simply to display whatever quantitative regularities may exist between the fundamental measurements made on bodies at each cosmic level.

The properties of the arithmetic relations between fundamental atomic and cosmic constants is not new ground. It has received the attention of many leading physicists and astronomers. Eddington (1923, 1931a,b); Haas (1930a,b, 1932, 1938a,b,c); Stewart (1931); Dirac (1937, 1938); Chandrasekhar (1937); Jordan (1937, 1947); Schrödinger (1938); Kothari (1938); Bondi (1952); Pegg (1968); Gamow (1968); and Alpher (1968) all have developed the subject.

The central theme in the numerical approach to atomic-cosmic relations has been to identify quantitative equivalences between various dimensionless combinations of fundamental constants and whenever possible give them physical interpretations. The epistemological weakness in this approach is the shadow of chance coincidence that cannot be removed by any of the common tests of statistical significance. Confidence in the validity of the numerically indicated relations can only follow from successful predictions or the development of a consistent theoretical construct linked to well established physics.

The basic ingredients in the relational approach are the micro-constants, e, m_e, m_p, and h (the charge and mass of the electron, the mass of the proton, and Planck's constant) the meso-constants, c and G (the velocity of light and the gravitational coupling constant), and the macroparameters H and ρ_u (the Hubble parameter and the mean density of the universe). Recently determined valves of these constants are given in Table I. From these fundamental quantities several important dimensionless ratios may be formed. The values of the dimensionless quantities $\mu = m_p/m_e$ (= 1836.12); $\alpha = 2\pi e^2/hc$ (= 1/137.0378); and $S = e^2/Gm_pm_e$ (= $10^{39.356}$) may

Table I

Values of Fundamental Physical and Cosmic Constants

Constant	Value (c.g.s.)	$\log_{10}$ (value)	Reference
e	4.80298×10^{-10}	−9.318489	1
m_e	9.10908×10^{-28}	−27.040526	1
m_p	1.67252×10^{-24}	−23.776629	1
h	6.62559×10^{-27}	−26.178776	1
c	2.997925×10^{10}	10.476821	1
G	6.670×10^{-8}	−7.176	1
H^{-1}	13×10^{9} years	17.613 seconds	2
ρ_u	10^{-28}	−28	3
a_o	5.29167×10^{-9}	−8.276407	1
r_e	2.81777×10^{-13}	−12.550095	1
α^{-1}	137.0388	2.136844	1
S	2.265×10^{38}	39.356	
μ	1836.12	3.263901	

From top: charge on electron, mass of electron, mass of proton, Planck's constant, velocity of light, Newton's gravitational constant, inverse Hubble parameter, mean density of visible matter in universe, Bohr radius, radius of electron, inverse fine structure constant, ratio of Coulomb to gravitational forces, ratio of proton to electron mass.

1. Cohen and DuMond (1965), 2. Sandage (1968) and 3. Allen (1963) p. 261.

be established in the laboratory. These are respectively, the ratio of proton to electron mass, the Sommerfeld fine structure constant, and the ratio of Coulomb to gravitational forces.[1]

When the two macro-parameters H and ρ_u are introduced, three additional dimensionless quantities may be formed. The first of these is the "scale parameter" of the universe (the product of the velocity of light, c, and the Hubble time H^{-1}), divided by the electron radius, c/Hr_e. The second is the "mass of the universe" expressed in units of baryon mass (where the scale parameter is taken as the radius of the universe), $\rho_u c^3/H^3 m_p$. The third is the dimensionless gravitational potential of the universe $GM_u/c^2 R_u = G\rho_u/H^2$. Using 75 km/sec/mpc as the present value of the Hubble parameter (Sandage 1968), and $10^{-28} g/cm^3$ for the mean density of matter in the universe (Allen 1963), we obtain:

$$c/Hr_e = 10^{40.64} \doteq 2\pi^2 S$$

$$G\rho_u/H^2 = 10^{0.05} \doteq 1.$$

$$\rho_u c^3/H^3 m_p = 10^{79} \doteq 2S^2$$

It is thus seen that to within small factors (whose exact value cannot be determined with the present precisions of ρ_u and H), the dimensionless cosmic quantities representing the potentital, size, and mass of the universe are closely equal to S^ν, where ν = 0, 1, and 2 respectively. The significant matter here is not the fact that the values differ from integral powers of S by factors

---

1 It has been recognized that S and α appear to be logarithmically related. As an example of an arithmetic equivalence presently lacking theoretical confirmation, we have $8\pi^2 S = 2^{1/\alpha}$ to within experimental uncertainties. If this equivalence is not a coincidence, it has several important implications. Bahcall and Schmidt (1967) have shown on the basis of 0 III emission pairs in the spectra of several radio galaxies with redshifts up to $\delta\lambda/\lambda = 0.2$ that α appears to have been constant for at least 2×10^9 years. The above equivalence, if non-coincidental, would imply that S has also been constant over this period. Hence if G has been changing with time, e^2 and/or m_p and m_e have also been changing, and if e^2 has been changing, so also has h and/or c. The gravitational constant may, indeed, be expressed in terms of other basic constants by the relation, $G = 8\pi^2 e^2/m_p m_e 2^{1/\alpha}$ (Wilson 1966).

as large as 2 or $2\pi^2$, but the fact that laboratory and observatory measurements of quite diverse phenomena when expressed in dimensionless form appear to approximate so closely some small power of the ratio of electric to gravitational forces. It is also interesting to note that the gravitational potential of the universe is near the Schwarzschild Limit, the theoretical maximum value for potential. These *quantitative* equivalences indicate that there probably exist basic causal *qualitative* relations between the structure of the universe and the properties of the atom and its nucleus (the question of the direction of causality being open).

So far the two levels represented by the atom and the universe as a whole have been shown to be derivable from integral powers of the basic dimensionless ratio S. Numerical relations of a similar type involving fractional powers of S were pointed out by Chandrasekhar (1937) to be related to other cosmic levels. Chandrasekhar formed the dimensional combination

$$M_\nu = \left(\frac{hc}{G}\right)^\nu m_p^{\,1-2\nu} \tag{1}$$

having the dimensions of mass. He pointed out the case $\nu = 3/2$ occurring in the theory of stellar interiors, leads to $M_{3/2} = 5.76 \times 10^{34}$ grams, the observed order of stellar masses. This is also the upper limit to the mass of completely degenerate configurations.

But the Chandrasekhar relation (1) also gives the observed order of mass for other cosmic levels in addition to the stellar level although this is not justifiable theoretically. If values of ν of the form $(2 - 1/n)$ where n is an even integer 2, 4, 6, 8, . . . are selected, then the Chandrasekhar relation predicts a sequence of masses given in Table II that corresponds to those observed for the stellar, galactic, cluster, second order cluster, ...levels of cosmic bodies.[2]

Table II. Masses for Levels of Cosmic Bodies from the Chandrasekhar Relation

Level	n	ν	$\log_{10} M_\nu$ (grams)	$\log_{10} M_\nu$ (dimensionless)
stellar	2	3/2	34.766	58.543
galactic	4	7/4	44.523	68.299
cluster	6	11/6	47.775	71.552
2° cluster	8	15/8	49.401	73.178
3° cluster	10	19/10	50.377	74.153
.....	...	.....	.....	.....
Universe	∞	2	54.280	78.056

Using well known relations between fundamental constants, equation (1) may be rewritten in the form:

$$M_\nu = \left(\frac{2\pi m_e}{\alpha m_p} S\right)^\nu m_p = A^\nu S^\nu m_p \tag{2}$$

where $A = 0.4689$. Hence the masses of the bodies on various cosmic levels defined by $\nu = 1\frac{1}{2}, 1\frac{3}{4}, 1\frac{5}{6}, 1\frac{7}{8}, \ldots, 2$, are seen to be nearly equal to these respective powers of S times the proton mass.

2. If equation (1) is valid for all ν of this sequence, then clusters of higher orders could exist until the ratio of consecutive cluster masses becomes less than two. The first pair for which this happens is $\nu = 31/16$ and $\nu = 35/18$, i.e., 6° and 7° clusters. Observationally, although 3° order clustering has been suspected (Wilson 1967), not even the existence of 2° order clustering has been satisfactorily established. While *even* values of n give masses in good agreement with cosmic levels, the *odd* values do not appear to correspond to any long lived objects. Nonetheless, if there exist two species of body, with masses $10^8 \odot$ and $10^{13} \odot$, such bodies would correspond to $n = 3$ and 5 respectively.

There are additional relations between the measurements of cosmic physics and microphysics. The largest gravitational potentials that have been observed for each of four species of cosmic bodies (stars, galaxies, clusters and 2° order clusters) are given in Table III. The potentials for each species are derived in physically distinct ways. For stars, from eclipsing binary observations; for galaxies, from rotational dynamics; for clusters, from the virial theorem; and for second order clusters, from angular diameters, distances and galaxy counts. It is interesting and somewhat surprising that the maximum in each case is nearly the same, a quantity of the order of 10^{23} grams/cm. If, instead of c.g.s. units, masses are expressed in baryon mass units and radii in Bohr radius units, the dimensionless ratio, $M/R \div m_p/a_o$, is in each case closely equal to 10^{39}. Thus, the upper bound for the gravitational potential of these species of cosmic bodies seems to be σS where σ is a factor of the order of unity not determinable from the present precision of the observational data.

Table III. Maximum Values of Potentials

System	$\log_{10}$ [M/R] (c.g.s.)	$\log_{10}$ [M/R] (dimensionless)
Stars	23.27	38.8
Galaxies	23.6	39.1
Clusters	23.5	39.0
Second-Order Clusters	23.2	38.7

From $M/R \leqslant \sigma S m_p/a_o$, substituting $e^2/Gm_p m_e$ for S and $e^2/m_e \alpha^2 c^2$ for a_o, we obtain

$$\frac{GM}{c^2 R} \leqslant \sigma \alpha^2$$

In other words, the dimensionless gravitational potential for these four species of cosmic bodies is bounded, not by the Schwarzschild limit, but by a bound α^2 times smaller. We thus see that not only the dimensionless microphysical quantity, S, but also the fine structure constant, α, emerges from cosmic measurement. (Another occurrence of α^2 in cosmic measurements derives from cluster redshifts (Wilson 1964).)

These results may be displayed graphically. Figure 1 is a small scale representation showing quantitative mass and size relations between atomic and cosmic bodies. The axes are logarithmic.

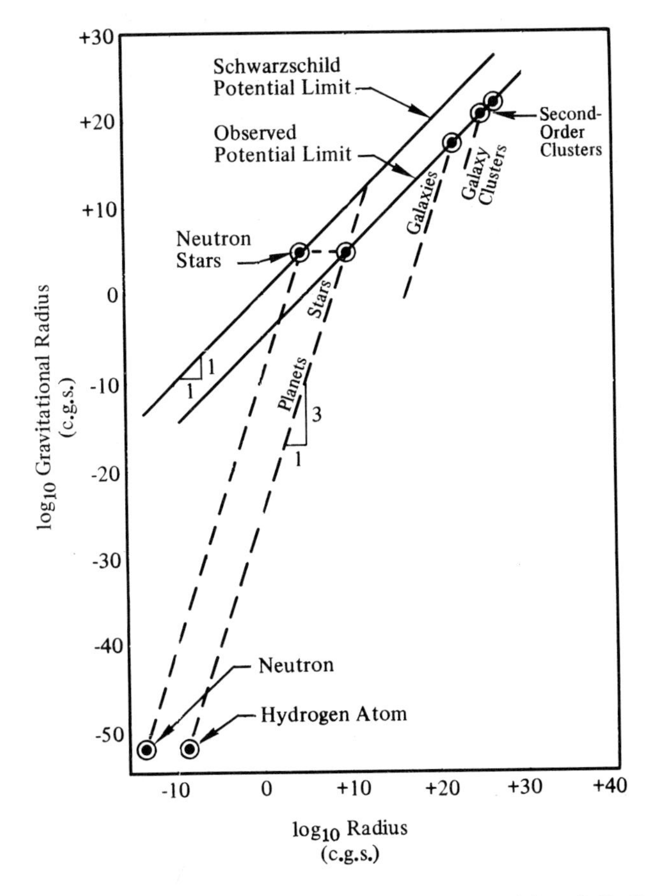

Figure 1 Mass and Size Relations Between Atomic and Cosmic Bodies

The abscissa represents the physical radius; the ordinate, the gravitational radius (GM/c^2). The upper 45 degree line is the Schwarzschild potential limit,

$$\frac{GM}{c^2 R} = \frac{1}{2},$$

the theoretical boundary separating the excluded region (upper left) from the allowable region for self-gravitating bodies. Such bodies as neutron stars, and presumably the universe itself lie on this limit. The lower 45 degree line is the observed or modular potential limit,

$$\frac{GM}{c^2 R} = \alpha^2,$$

marking the locations of the various cosmic bodies having the maximum observed potentials. All other stars, galaxies, clusters, etc., lie below this limit. The relation of the nucleus of the atom and the atom to the degenerate neutron star and the normal star is shown by the dotted lines of constant density (slope 3). Thus a neutron star has the largest mass with nuclear density allowed by the Schwarzschild limit. A normal main sequence star is seen to be limited to the same mass but is non-degenerate, lying on the line representing "atomic density." Thus, given the properties of the atom and the Schwarzschild limit, it is possible to derive the observed maximum mass for a star, but as with the Chandrasekhar relation, it is difficult to account for the locations on the diagram of the bodies of lower density (clusters, galaxies, etc.) and the fact that they are also bounded by the α^2 potential limit.

The parallel lines of equal density (slope 3) through the atom, planets and normal stars, the star clusters and galaxies, the clusters, etc., represent the levels of a modular hierarchy as previously described. These levels are thus definable by a discrete density parameter. Further, in consequence of the universal relation for gravitating systems, $\tau \propto \rho^{-1/2}$, relating a characteristic time to the density, the levels in the cosmic modular hierarchy are also definable in terms of a discrete *time* or *frequency* parameter. We shall return to this concept later.

MASS BOUNDS

In order to display the cosmic or upper portion of Figure 1 with more detail and to make comparisons with observations, the logarithms of observed masses (M) and potentials (M/R) of planets, stars, globular star clusters, galaxies, and clusters of galaxies have been plotted in Figure 2. The masses and potentials (Allen 1963) include maximum and minimum observed values and other representative values selected to show the domains occupied by the respective cosmic species.

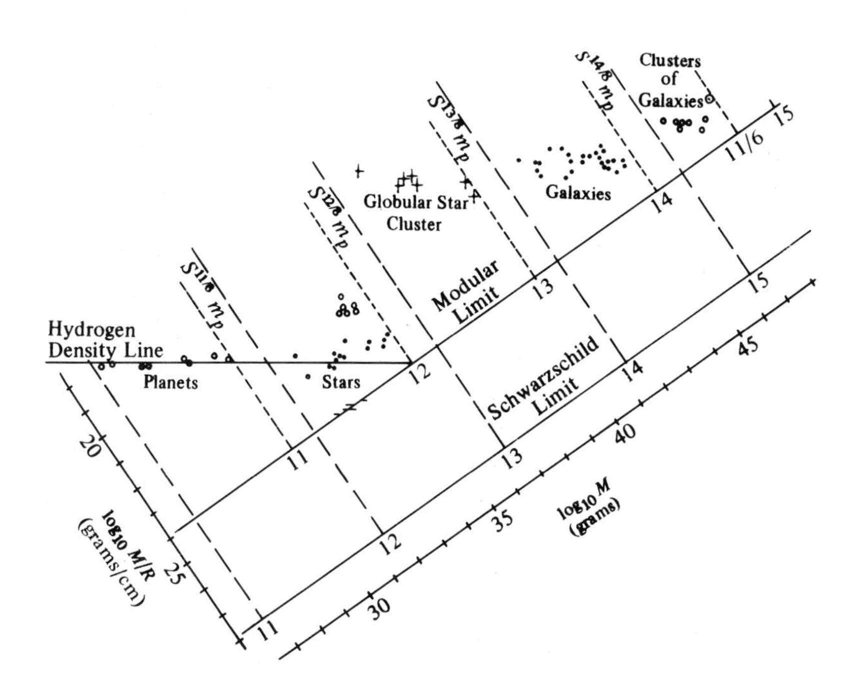

Figure 2 Mass Bounds of Cosmic Bodies

However, because of observational bias toward brightest and largest objects, the minimum observed values are not as representative of actual minimum values as the maximum observed values are of actual maximum values. Figure 2 is related to Figure 1 by an affine transformation (Figure 1 has not only been dialated, but has also been subjected to shear, reflection and rotation transformations). In Figure 2, the lines of constant density are shown horizontally so as to display the levels into which cosmic bodies fall when viewed as a modular hierarchy.

The supergiant stars lying above the mean stellar density level are shown as open circles, while the white dwarfs lying below the level near the modular potential limit are shown as dashes. The Schwarzschild Limit, $M/R = c^2/2G$ and the modular (or observed) limit, $M/R = Sm_p/a_o$ have a slope of 2/3 with respect to the horizontal equi-density lines. The short-dashed and long-dashed lines perpendicular to the Schwarzschild and modular limits are lines of constant mass. The set of short-dashed lines, extending only to the modular limit represent the sequence of masses $M_\nu = S^\nu m_p$, showing values of ν = 11/8, 12/8, 13/8, 14/8, and 11/6. The set of long-dashed mass lines, extending to the Schwarzschild Limit are located so as to pass through a sequence of points on the Schwarzschild Limit that have the same gravitational energy as the intersections of the $S^\nu m_p$ mass lines with modular limit. The pairs of intersections marked 14, 13, 12, . . .lie on lines of constant gravitational energy, $GM^2/R = S^\nu m_p(\alpha c)^2$. For identification, corresponding upper and lower bound intersections with the modular and the Schwarzschild Limits are marked with the *numerators* of the exponent ν. That is, 14 on the Schwarzschild Limit marks the lower bound of galaxies and corresponds to the upper bound $S^{14/8} m_p$ intersection with the modular limit.

The values of mass given by the Chandrasekhar relation (1) in Table II are the correct order of magnitude for the masses of stars, galaxies, and clusters. In Figure 2 it can be seen from the set of short-dashed lines of constant mass that the sequence of masses $S^{\nu}m_p$ are close in value to least upper bounds of the masses of planets, stars, globular star clusters, galaxies, and clusters of galaxies. Numerical comparisons of maxima are given in Table IV. In addition, the set of long-dashed lines are seen to be lower bounds, while probably not greatest lower bounds nonetheless close to the actual observed minimum values of the masses of the respective species of cosmic bodies. Numerical comparisons of minima are also given in Table IV where the lower bounds are the upper bounds diminished by $10^{3.9}m_p$. It can be shown that this value of maximum-minimum mass differential may be derived from "ν sequences" of maximum

Table IV. Observed and Calculated Mass Limits

Mass Limit	Planets	Stars	Globular Clusters	Galaxies	Galaxy Clusters
MAXIMUM					
	Jupiter	VV Cephei A	M22	M31	Local Super Cluster
Observed	30.279	35.225	40.14	44.8	48.3
Model	30.338	35.258	40.176	45.096	48.376
$S^{\nu}m_p$	$\nu = 11/8$	$\nu = 12/8$	$\nu = 13/8$	$\nu = 14/8$	$\nu = 11/6$
MINIMUM					
	Mercury	R CMa B	M5	NGC6822	
Observed	26.509	32.340	37.3	41.9	
Model	26.4	31.4	36.3	41.2	

All masses are given in $\log_{10}$ (grams). Upper bounds are given by $S^{\nu}m_p$, lower bounds by $S^{\nu}10^{-3.9}m_p$.

masses and gravitational energies, with the minimum mass being the least allowed by the Schwarzschild Limit for a given gravitational energy.

THE COSMIC DIAGRAM

The good agreement between the observed values for the masses and sizes of various species of cosmic bodies and the values given by sequences involving simple expressions containing fundamental physical constants indicates the probable validity of the gross features of the sequences. However, systematic errors and incompleteness in the observational data and the uncertainties intrinsic in establishing observationally least upper bounds and greatest lower bounds render it impossible, in the absence of a rigorous physical theory, to predict the exact form of the expressions and the values of the small factors (such as the 2π's, etc.) that should be included. We might, as an analogy, think of our discerning Kepler's Third Law in the form: periods squared are proportional to orbital diameters cubed without knowing the important constant of proportionality, $G(M_1 + M_2)$.

In the spirit of focusing on the major patterns that emerge from the present body of observations that are not likely to be seriously altered by refinements in observation, or even by discovery of new bodies, we represent the gross features of the structure in the universe in Figure 3. In this stylized representation, the cosmos is mapped on a rectangle whose length is the logarithm of the mass, $S^{\nu} m_p$, and whose hieght is the logarithm of the extension, $S^{\eta} a_o$. The masses and radii of various sub-components are related to values of ν and η. The hydrogen atom, mass m_p, and radius a_o, is located at the origin at H with $\nu = 0$, $\eta = 0$. The mass and radius of the universe are represented by the values $\nu = 2$, $\eta = 1$ at U. The modular and Schwarzschild potential limits are the upper and lower 45° lines respectively. The remaining observed bodies in the universe lie roughly within the three hatched bands, whose slope is that of constant density terminating at the modular limit. The bodies on the lowest and longest band have density of the order of one g/cm^3 and include asteroids, satellites, planets, and stars. This band terminates on the modular limit at $\nu = 3/2, \eta = 1/2$. With little mass overlap of the first sequence, the next sequence of bodies (star clusters and galaxies) begins near $\nu = 3/2$ and falls along an equi-density band reaching the modular limit at $\nu = 7/4$, $\eta = 3/4$. Above this point the observational uncertainties do not permit a definitive picture. It is not clear whether there exist two (or more) sequences of clusters of galaxies or only one.

A cluster sequence terminating at $\nu = 11/6, \eta = 5/6$ together with a second sequence of higher order clusters terminating at $\nu = 15/8$, $\eta = 7/8$ (as shown in Figure 1 and Figure 2) may fit observations better than the single sequence extending to $\nu = 15/8$, $\eta = 7/8$ shown in Figure 3. The resolution of this structure as well as whether still higher levels of clustering exist must be decided on the basis of future observations.

From the point of view of hierarchies, the levels occupied by cosmic bodies may be described either as *modular levels* (in the sense defined earlier), or as levels defined by a density

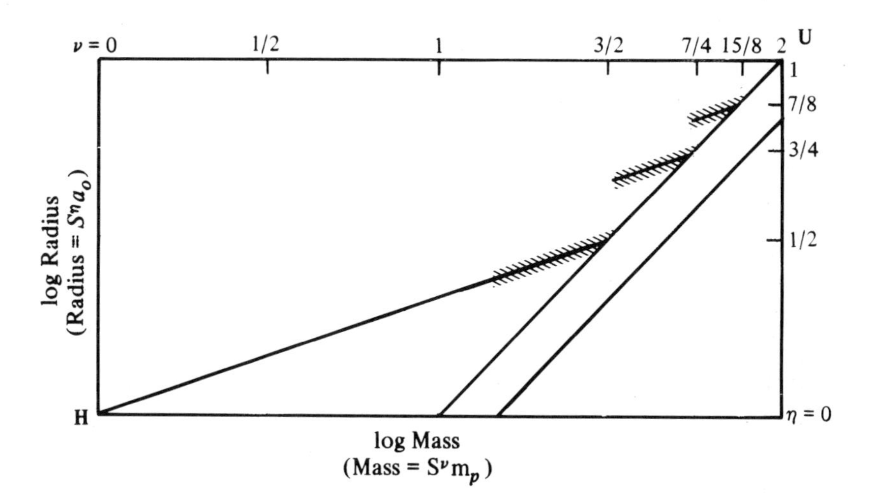

Figure 3 Cosmic Diagram

parameter, or its equivalent frequency parameter. In addition the structure may be "sliced" differently and the cosmic bodies may be allotted to distinct levels defined by a mass parameter. These levels are broad but on the scale of Figure 2 appear to be distinct.

INTERPRETATIONS

An intrinsic difficulty in relating empirical results (such as those displayed in Figures 2 and 3) to current physical theories is that numbers of the magnitude of S are not contained in any classical equations of physics. This difficulty has been expounded by Dirac (1938), Jordan (1947) and others. Eddington (1931) made attempts to derive the fundamental dimensionless constants from first principles, not, however, with complete success in reproducing the observed values. A theoretical understanding of the various observed relations between the different levels of cosmic structure – atoms, stars, galaxies, . . .the universe – is thus likely to come only after new theories of such concepts as time, degeneracy, and informational content of structure are available. At the present stage only some *speculative* suggestions can be made.

For example, the existence of *two* potential limits, the Schwarzschild and the modular, implying that the same extension ratio (the α^2 ratio of atomic to nuclear dimensions) holds between non-degenerate and collapsed configurations at stellar, galactic and cluster levels, suggests that through a generalization of the concept of degeneracy, the theorectical validity of equation (1) for all levels might be established. One might speculate that configurations at every level possess a collapsed or close packed state, and an extended state α^{-2} times larger. An alternate approach may be that the reflection of the α^2 ratio into higher levels of cosmic-structure is a cosmogonic vestige from a universe in a highly collapsed state. But whatever the cause of the modular limit, it must be regarded as an important observational feature to be accounted for by cosmological theories.

A second speculative suggestion is that in the sequence of powers of S that map observed mass configurations, we are encountering a resonance phenomenon. However, the fundamental and the overtones are exponentially related instead of being related in the manner of Pythagorean harmonics. This suggests kinship to the logarithmic time derived by Milne (1935) in his kinematic relativity. If we take as the basic gravitational frequency, the inverse Schuster period, $f_o = (Gm_p)^{1/2}/2\pi a_o{}^{3/2}$, then the overtones are given by

$$f_\nu = \frac{(GS^\nu m_p)^{1/2}}{2\pi(S^{\nu-1}a_o)^{3/2}} = f_o S^{3/2-\nu} \qquad (3)$$

where $\nu = 3/2, 7/4, 15/8, \ldots.$

Numerically, $f_{3/2} = f_o$, the frequency associated with the hydrogen-stellar line of Figure 3, corresponds to a period of about two hours; $f_{7/4}$, the galactic line corresponds to 10^6 years; $f_{15/8}$, the cluster line corresponds to 85 x 10^9 years; and f_2 corresponds to 10^{15} years. The cluster value is close to the period derived by Sandage for an oscillating universe. Viewed as a Hubble time, it corresponds to a value of H = 74.13 km/sec/mpc, in close agreement with the observed value of H = 75.3 km/sec/mpc derived from cluster distances (Sandage 1968).

If we take this equivalence between the ν = 15/8 cluster gravitational time and the observed cluster Hubble time, as additional corroboration of the valid representation of the cosmic diagram, then we infer that the visible sample of the universe, the "realm of the galaxies and clusters" is not the ν = 2 universe. The observations at the limits of our telescopes are describing the ν = 15/8 sub-structure and not the universe. Characteristic times of the order of 10^{10} years are those associated with the cluster level sub-structure. The characteristic gravitational time of the ν = 2 universe, on the other hand, is of the order of 10^{15} years. The appearance of a time of this magnitude brings to mind the controversy that waged in cosmology following the publication of James Jeans (1929) estimate of the dynamic age of the galaxy at 10^{13} years. The adherents of the "short time-scale," held the age of the universe to be but a few eons while those who subscribed to the "long time-scale," required an age of the order of 10^{13} years or greater. Since the galaxy could not be older than the universe, the issue was settled against Jeans. But if the few eons refers not to the universe but to the cluster level sub-structure, there is no *a priori* reason why the galaxy cannot be older than the cluster level sub-structure.

If the cosmic diagram suggests some form of resonance as the process of morphogenesis, then as sand collects at the nodes on a vibrating drum head, matter concentrates at nodes corresponding to the set of frequencies $S^{3/2-\nu} f_o$. This raises many physical questions. Most importantly, what is it that is pulsating or vibrating at these frequencies – some substratum, matter itself, or what? Analogies to familiar equations suggest that from the cosmic diagram, we have a set of eigen values representing mass levels, energy levels, or frequencies that are solutions to some "cosmic wave equation." Perhaps the first step toward a physical theory would be to derive such an equation.

REFERENCES

Allen, C. W. 1963. *Astrophysical Quantities.* 2nd ed. London: The Athlone Press.

Alpher, R. A., and Gamow, G. 1968. "A Possible Relation Between Cosmological Quantities and the Characteristics of Elementary Particles." *Proc. Nat. Acad. Sci* 61:363.

Bahcall, J. N., and Schmidt, M. 1967. "Does the Fine-Structure Constant Vary with Cosmic Time?" *Phys. Rev. Letters* 19:1294.

Bondi, H. 1952. *Cosmology.* Cambridge: Cambridge University Press.

Chandrasekhar, S. 1937. "The Cosmological Constants." *Nature* 139:757.

Charlier, C. V. L. 1908. "Wie eine unendliche Welt aufgebaut sein kann." *Arkiv för Matematik, Astronomi och Fysik.* Band 4, No. 24.

——. 1922. "How an Infinite World May be Built Up." *Arkiv för Matematik, Astronomi och Fysik.* Band 16. No. 22, pp. 1-34.

Cohen, E. R., and DuMond, J. W. N. 1965. "Our Knowledge of the Fundamental Constants of Physics and Chemistry in 1965." *Phys. Rev.* 37:537.

Dirac, P. A. M. 1937. "The Cosmological Constant." *Nature* 139:323.

——. 1938. "A New Basis for Cosmology." *Proc. Roy. Soc., Series A* 165:199.

Eddington, A. S. 1923. *The Mathematical Theory of Relativity.* Cambridge: Cambridge University Press.

——. 1931a. "Preliminary Note on the Masses of the Electron, the Proton, and the Universe." *Proc. Cambridge, Phil. Soc.* 27:15.

——. 1931b. "On the Value of the Cosmical Constant." *Proc. Roy. Soc., Series A.* 133:605.

Gamow, G. 1968. "Numerology of the Constants of Nature." *Proc. Nat. Acad. Sci.* 59:313.

Haas, A. E. 1930a. "Die mittlere Massendichte des Universums." *Anzeiger der Akad. Wiss. Wien.* 67:159.

——. 1930b. "über den möglichen Zusammenhang zwischen kosmischen und physikalischen Konstanten." *Anzeiger der Akad. Wiss. Wien.* 67:161.

——. 1932. "über die Beziehung zwischen Krummungsradius der Welt und Elektronenradius." *Anzeiger der Akad. Wiss. Wien.* 67:91.

——. 1938a. "A Relation Between the Average Mass of the Fixed Stars and the Cosmic Constants." *Science* 87:195.

——. 1938b. "A Relation Between the Electronic Radius and the Compton Wavelength of the Proton." *Science* 87:584.

——. 1938c. "The Dimensionless Constants of Physics." *Proc. Nat. Acad. Sci.* 24:274.

Jeans, J. 1929. *Astronomy and Cosmogony.* Cambridge: Cambridge University Press.

Jordan, P. 1937. "Die Physikalischen Weltkonstanten." *Die Naturwissenscaften* 25:513.

——. 1947. *Herkunft der Sterne.* Stuttgart: Wiss, Verlag.

Kothari, D. S. 1938. "Cosmological and Atomic Constants." *Nature* 142:354.

Lambert, J. H. 1761. *Kosmologische Brieffe.* Leipzig.

Milne, E. A. 1935. *Reliability, Gravitation and World Structure.* Oxford: Oxford University Press.

Pegg, D. T. 1968. "Cosmology and Electrodynamics." *Nature* 220:154.

Sandage, A. R. 1968. "Directors Report, Mt. Wilson and Palomar Observatories, 1967-1968." p. 33. Pasadena, California.

Schrödinger, E. 1938. "Mean Free Path of Protons in the Universe." *Nature* 141:410.

Shapley, H. 1930. *Flights from Chaos.* New York: McGraw-Hill.

Stewart, J. Q. 1931. "Nebular Red Shift and Universal Constants." *Phys. Rev.* 38:2071.

Wilson, A. G. 1964. "Discretized Structure in the Distribution of Clusters of Galaxies." *Proc. Nat. Acad. Sci.* 52:847-54.

——. 1966. "Dimensionless Physical Constants in Terms of Mathematical Constants." *Nature* 212:862.

——. 1967. "A Hierarchical Cosmological Model." *Astronom. J.* 72:326.

——. 1969. "Closure, Entity and Level" (this volume).

■

Dimension as Level

Robert Edward Williams*

In geometry, dimensionality is usually conceived as a set of independent, mutually perpendicular space coordinates $x, y, z, \ldots$. Dimension is not ordinarily considered to be a 'level' or to be associated with an 'emergent whole' of higher order that arises from lower order elements (Bunge 1960). It is the purpose of this note to point out a sense in which dimension can be considered to be a level, and give an illustration of emergence using geometrical polytopes.

Eighteenth century geometers realized (Coxeter 1963) that a certain algebraic sum of realizable geometrical entities in a polytope is either equal to zero for $n = 0, 2, 4, \ldots$, dimensions or equal to two for $n = 1, 3, 5, \ldots$, dimensions. This relation known as Euler's Law (Euler 1752) was generalized to n dimensions by Schläfli in 1852 (Coxeter 1968) and proved by Poincaré (1893). This law can be written:

$$N_0 - N_1 + N_2 - \ldots + (-1)^{n-1} N_{n-1} = 1 - (-1)^n, \tag{1}$$

where N is the number of entities and the subscript is dimension. Specifically, N_0 is the number of vertices, N_1, the number of edges, N_2, the number of faces, N_3, the number of solids, ...

To appreciate how a new level emerges from combining entities of lower dimension, consider the case of equation (1), for a two-dimensional tessellation,

$$N_0 - N_1 + N_2 = 1. \tag{2}$$

Combining polygons under the rule that (i) we maintain Euclidian space, (ii) we do not distort the polygons, and (iii) that every two polygons share one edge, we may build up an indefinitly large aggregate of connected polygons satisfying equation (2). If we are aggregating pentagons, for example, when we accumulate twelve (Figure 1a) and connect them so as to join *all* edges, we obtain a dodecahedron (Figure 1b).

Douglas Advanced Research Laboratories, Huntington Beach, California. Current address: Department of Design, University of Sourthern Illinois, Carbondale, Illinois 62901.

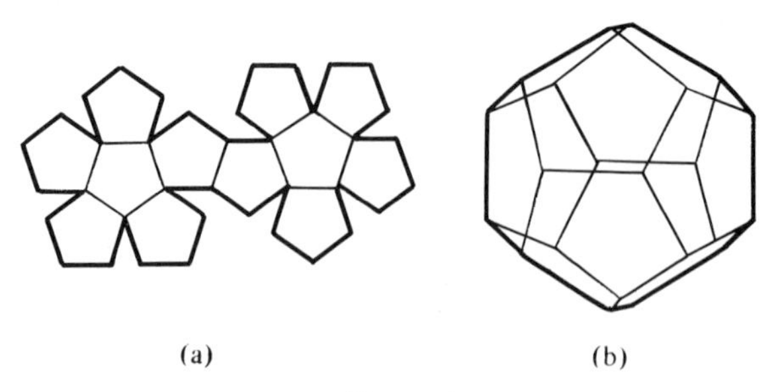

(a) (b)

Figure 1

The above conditions governing the combination rule are not violated, but the resulting single entity no longer satisfies equation (2). Instead,

$$N_0 - N_1 + N_2 = 2 \tag{3}$$

where the whole in the right member is greater than the old sum of parts. Euler's Law can be restored by introducing N_3, a new level, a new dimension and an emergent entity.

In like manner, an entity of dimension $n = 4$ emerges from combining entities of dimension $n = 3$ according to our combining rule: It is also possible to achieve emergence of a new dimension either by distorting the elements (Figure 2) or by creating a suitably curved space.

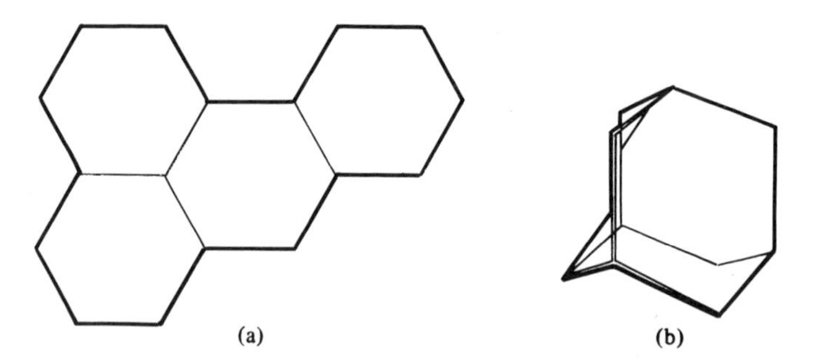

(a) (b)

Figure 2

We have illustrated special cases of Euler's Law (for n dimensions) with a combining rule of three conditions. An infinite number of polytopes may be generated from either Euclidean or non-Euclidean entities of dimension $(n - 1)$. Whether newly formed polytopes are regular or not can be determined by known rules. Dimension $(n + 1)$ emerges from the operation of combining entities of *n* dimension. Therefore, the concepts of emergence and level are fundamental in the geometry of aggregating regular polytopes.

REFERENCES

Bunge, M. 1960. "Levels: A Semantical Preliminary:" *Review of Metaphysics* 13: 396–406.

Coxeter, H. S. M. 1963. *Regular Polytopes,* 2nd ed. p. 165. New York: Macmillan.

——. 1968. Twelve Geometrical Essays, p. 233. Carbondale, Ill.: Southern Illinois University Press.

Euler, L. 1752. "Elementa doctrinae solidorum," *Novi commentarii Academiae scientiarum imperialis petropolitanae* 4: 109–160.

Poincaré, H. 1893. "Sur la généralisation d'un théorème d'Euler relatif aux polyèdres." *Comptes rendus hebdomadaires des séances de l'Académie des Sciences, Paris* 117: 144–45.

■

Overlap in Hierarchical Structures

Paul J. Shlichta*

Most analyses of hierarchical structures can be represented by tree-like diagrams wherein each element of the system belongs to only one subsystem at each hierarchical level. There is an understandable aesthetic preference for such systems and a corresponding aversion to systems having elements which, at a given level, belong to more than one subsystem and which therefore are connected to the highest level by more than one hierarchical path. This prejudice, however, has recently been challenged on the grounds that a healthy urban milieu normally involves a high degree of organizational overlap (Alexander 1965). In corroboration of this, the present note demonstates that even in the most abstract and in rigidly defined systems, horizontal overlap is prevelent. Three examples – from group theory, geometry, and crystallography – are briefly cited. A more detailed exposition is presented elsewhere (Shlichta 1969).

SYMMETRY GROUPS

The existence and interrelation of intermediate entities in a set of elements are nowhere more rigidly defined than in the theory of finite groups. A set of elements constitutes a group only if certain criteria are satisfied. If these elements also belong to a larger group, then the smaller group is a subgroup of the larger one (Margeneau and Murphy 1943). Therefore a system of groups has all the attributes of a hierarchical structure – one in which, however, a high degree of overlap is almost invariably encountered. The crystallographic point groups – the sets of rotations, mirror reflections, and inversion centers which define the symmetry of finite bodies – provide a typical example. Their hierarchical diagram, as shown in Figure 1 (after Neubuser and Wondratscheck 1966), consists of two partially overlapping trees, each of which has considerable internal overlap. This structure, although theoretically valid, is so complex and aesthetically repugnant that it has little didactic or mnemonic value and other classification schemes are generally preferred (Henry and Lonsdale 1965).

POLYHEDRA

A polyhedron may be defined as a three-dimensional cell bounded by polygonal faces, each of which is bounded by edges which are in turn bounded by vertices. Since the lower-dimensional boundaries are both subsystems and components of the higher-dimensional ones, a polyhedron is a perfect example of a hierarchical structure–one in which

**Douglas Advanced Research Laboratories, Huntington Beach, California, 92647.*

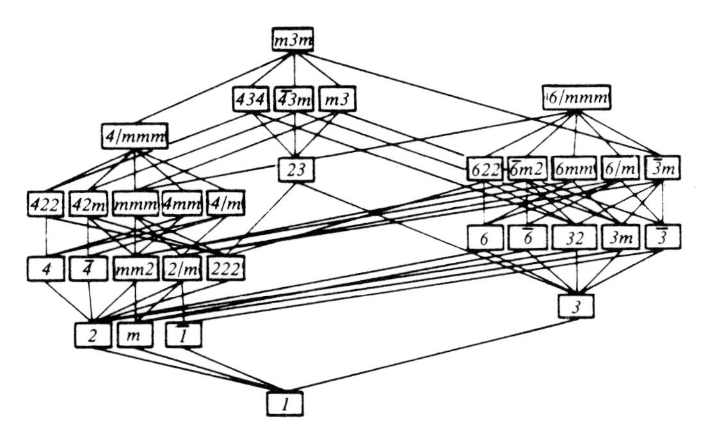

Figure 1

dimensionality is an index of level (Williams 1969). In such systems, however, two kinds of overlap are encountered: (i) although the dissection of a polyhedron into its faces is non-overlapping, the further dissection of each face entails overlap since each edge is shared by two faces and each vertex by three or more faces; this is evident in the lower levels of Figure 2. (ii) It is often necessary to recognize the existence of additional overlapping levels corresponding to aggregates of faces. The validity of these aggregates and entities can be justified by a geometric formulation of closure (Shlichta 1969). The most common example is the coordination polygon – the ring of faces sharing a common vertex. On an icosohedral sphere, for example, each vertex is the center of a pentagon of five faces (Figure 3a). Since each face is shared by three such pentagons (Figure 3b), the inclusion of these domains greatly increases the overlap and complexity of the hierarchical diagram of an icosohedron (Figure 2). This proliferation of overlapping levels becomes much worse in hyperdimensional systems. For example, the regular polytope $\{3,3,5\}$ in which a finite curved space is partitioned into 600 regular tetrahedra, has many levels of closure-justifiable aggregates including 5-cell bipyramids (Figure 3c) wherein each cell is shared by six clusters, 20-cell icosohedra (Figure 3d) wherein each cell is shared by four clusters, and an innumerable series of five-fold columns (Figure 3e). The construction of the complete hierarchical diagram of this system is left to the reader.

CRYSTAL STRUCTURES

A crystal possesses hierarchical structure inasmuch as it is made up of atoms which in turn are composed of electrons and nuclei that have complex infrastructures of their own. Moreover, since the crystal is held

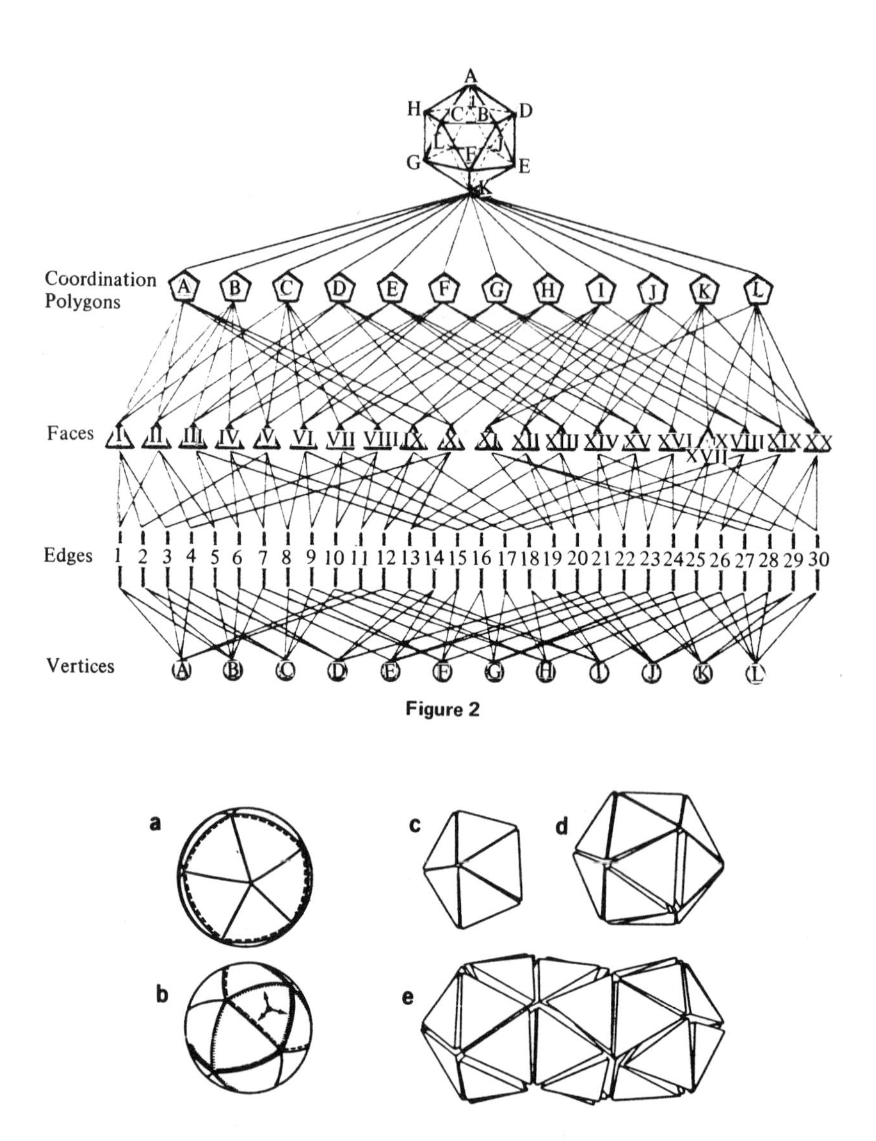

Figure 2

Figure 3

together by some sort of sharing of electrons between atoms, there is obviously considerable overlap at this level. Here, however, we are concerned with the existence of levels between the crystal and the individual atoms – that is, clusters of atoms. This is most obvious in organic crystals in which atoms are bonded together into molecules which in turn are packed to form the crystal. In such systems, electronic overlap is only a second-order effect. A much more significant kind of overlapping structure is evident in the classification and interpretation of metallic or ionic crystals. These crystals can of course be divided into non-overlapping atomic regions by means of Dirichlet regions (Coxeter 1961) or by nodal polyhedra which correspond to the topological bonding network of the crystal (Burt (1966; Schoen 1967; and Shlichta 1967). These models, however, are of little value in classifying structures since they do not reveal similiarities which exist between closely related and interconvertable structures such as face-centered and body-centered cubic (Figures 4a and 4b). On the other hand, the first coordination shells of these two structures shown in Figure 4c clearly show that one is a slight affine distortion of the other. Similarly, most successful schemes for classifying and predicting intermetallic structures are based on some form of atomic cluster in which any one atom is shared by several clusters (Frank and Kasper 1958; Pauling, pp. 425-31, 1967; and Samson 1967). Analyses of ionic crystals such as the rock-forming silicate minerals also employ coordination polyhedra which share vertices, edges, and faces (Pauling, pp. 544-64, 1967), If successful classification and prediction is a criterion of validity, crystal structures may be said to have as high a degree of overlap as the mathematical examples described above.

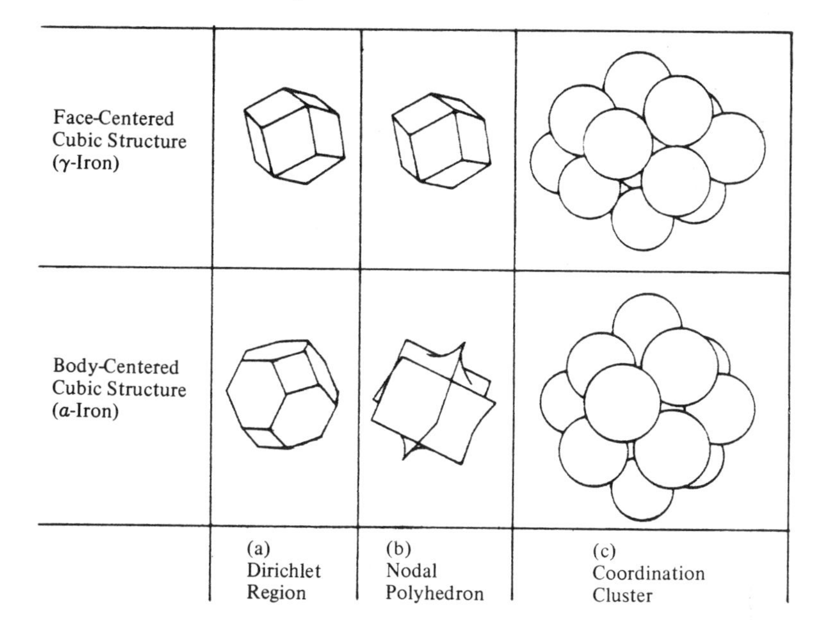

Figure 4

REQUIREMENTS FOR AVOIDING OVERLAP

It would be foolhardy to attempt to extrapolate from these comparatively simple and exact structures to the complexities and subtilties of biological and human systems. However, a few generalizations are perhaps not out of place. First, it is obvious that there are many possible causes of overlap; different ones are operative in each of the examples cited. Therefore, rather than attempting to enumerate all the causes of overlap, let us consider the stringent requirements for avoiding it. These are in part:

1) *Subordination:* the forces binding elements to the first level subsystem must be weaker than the forces joining the subsystems into larger systems; otherwise either the subsystems will tend to merge and share elements or additional, overlapping subsystems will emerge.

2) *Localization:* the forces attracting individual elements to each other must be so localized or so specialized that, once a subsystem is formed, there is no appreciable attraction or interaction between elements in different subsystems.

3) *Monodimensionality:* every element and subsystem must be susceptible to only one principle of aggregation so that there is only one way in which they can combine to form larger systems. If the overall system has a plurality of functions, interests, or aspects then a multidimensional type of overlap is inevitable.

The requirements for avoiding overlap are so restrictive that there are very few natural systems that can fit them. It appears to be much easier – and healthier – to learn to live with overlap than to avoid it.

In this connection, a final word might be said about the use of tree-like diagrams to study hierarchical structures. It is obvious that as the degree of overlap increases, the usefulness of such diagrams diminishes. Figure 2 may be regarded as a *reductio ad absurdam;* the icosohedron, considered of itself, is much easier to comprehend than its hierarchical diagram. One might suggest, therefore, that some other means of hierarchical analysis would be preferable. For example, this note has shown that a relation between closure and hierarchical structure exists in mathematical groups and polyhedra; it may be that closure can be used as a primary criterion for the existence of hierarchical levels in general (Wilson 1969).

REFERENCES

Alexander, C. 1965. "A City is Not a Tree." *Architectural Forum.* Reprinted in *Design,* No. 206, February, 1966. pp. 46-55.

Burt, M. 1966. "Spatial Arrangement and Polyhedra with Curved Surfaces and Their Architectural Applications." Masters Thesis, Israel Institute of Technology, Haifa.

Coxeter, H. S. M. 1961. *Introduction to Geometry.* New York: John Wiley & Sons. (The original reference to Dirichlet regions is Dirichlet, G. L. 1850. *Journal für reine und angewandte Mathematik* 40:216-19.)

Frank, F. C., and Kaspar, J. S. 1958. "Complex Alloy Structures Regarded as Sphere Packings." *Acta Cryst.* 11:184-90.

Henry, N. F. M., and Lonsdale, K. 1965. *International Tables for X-Ray Crystallography: Vol. I. Symmetry Groups,* pp. 22-29. 2nd ed. Birmingham, Eng.: Kynock.

Margenau, H., and Murphy, G. M. 1943. *Mathematics of Physics and Chemistry,* p.526. New York: Van Nostrand.

Neubuser, J., and Wondratschek, H. 1966. "Untergruppen der Raumgruppen." *Krist. Tech.* 1:529-543.

Pauling, L. 1960. *The Nature of the Chemical Bond.* 3rd ed. Ithaca, N.Y.: Cornell University Press.

Samson, S. 1967. "The Crystal Structure of the Intermetallic Compound $Cu_4 Cd_3$." *Acta Cryst.* 23:586-600.

Schoen, A. 1967. "Homogeneous Nets and Their Fundamental Regions." *Notices of the Amer. Math. Soc.* 14:661.

Shlichta, P. J. 1967. "A Topological Approach to Polyhedra Packing and Structural Networks." *Amer. Crystallographic Assoc. – Program and Abstracts: Minneapolis Meeting* p.55.

——. 1969. *Some Mathematical and Crystallographic Examples of Overlap in Hierarchical Structures.* Research Communication No. 111, Douglas Advanced Research Laboratories, Huntington Beach, Calif.

Williams, R. E. 1969. "Dimension as Level" (this volume).

Wilson, A. G. 1969. "Closure, Entity and Level" (this volume).

■

Part III

Organic Hierarchical Structures

It is in the organic world that the concepts of level and hierarchy acquire added dimensionality and greater subtlety. While the structure of the inorganic world is such that models that do not employ levels, hierarchies, or vertical relations have been extremely successful in explaining and predicting phenomena, the basic properties of the organic world require models whose construction quickly leads into matters of levels and their relation to one another. Part III begins with an historical account by Chauncey Leake of the development of awareness of organic levels in man and nature. Starting with the differentiation between the individual and the group in prehistoric times, Leake traces the discovery of organs, cells, and sub-cellular units, leading to the recognition in modern times of the molecular level at the lower end of the organic hierarchy and the ecological level at the upper end.

Howard Pattee discusses some reductionist-holist aspects of hierarchical models of bio-organisms. From the perspective of the upper levels, controling constraints are taken for granted and the problem is to explain how the organism works. From the perspective of the lower levels with elements that obey the laws of physics, the problem is to show how the constraints that control the elements arise from a collection of elements and generate an integrated function or purpose. Pattee distinguishes structural, functional, and descriptive hierarchies and concludes that all hierarchical organizations require a balance between the number of degrees of freedom of their elements, the fixed constraints that function as a record, and flexible constraints that control.

Robert Rosen joins Pattee in pointing out that appropriate to every level of a hierarchy, there is a different system description or language. He goes on to formulate the problem of levels as the phenomenological specifying of macrosystem behavior in terms of suitable observables, the specification of microsystem dynamics, and the development of a formalism (like statistical mechanics) connecting the two. He concludes that hierarchical structures cannot be based solely on automata-theoretic descriptions since the mechanism to generate higher level descriptions is explicitly abstracted out of the description at the outset.

John Platt illustrates the importance of the concept of boundaries to hierarchical systems and develops several of their functional attributes: boundary coincidence for different properties making a "thing" perceivable; gradients and flows being parallel or perpendicular to boundaries; ratios of interconnections to gates (spatial and temporal, distributed or concentrated) limiting the ability of the system to sense and respond to the external world.

In summarizing the symposium, Ralph Gerard reemphasized the roles of boundaries and edges, gradients, integration, and function. The more highly integrated an organism, the larger are the forces operating down with respect to the forces acting upward. Evolution of systems is toward higher integration with an increase in the number of levels. With structure (the system component constant in time) and function (the reversible behavior) and evolution (the irreversable behavior) there is an evolutionary spiral or helix of structure determining function and function producing structure. Gerard also feels it is premature to differentiate "system" and "hierarchy", but it is most important to order by origin as well as to order by function. In the note that follows, Herbert Gutman outlines the argument that an understanding of the genesis of hierarchies in living systems must proceed from a fundamental clarification of the relationship of structure to function and of organic wholes to their parts.

■

Historical Aspects of the Concept of Organizational Levels of Living Material

Chauncey D. Leake*

Recognition of organizational levels of living material has come with powerful impact in recent decades both scientifically and theoretically, in relation to molecular biology at a basic level, and popularly, thanks to Rachel Carson, at the far reaching ecological level. Historically, the development of the concept of organizational levels of living material offers an interesting example of intellectual evolution, moving outward in all directions from ourselves, ontogenetically and phylogenetically centralized to our environment.

IN THE BEGINNING: INDIVIDUALS AND SOCIETIES

Implicit in the earliest family experience in remote antiquity is a realization, conscious or unconscious, of the distinction between an individual and the family group to which that individual might belong. This would have occurred either in a natural or an adoptive family grouping, and the distinction would have been apparent to all who comprised the family. This dichotomy would have sharpened with tribal development. It was the major factor in the origin of at least three ethical principles. These ethics are the earliest answers to that basic question which still concerns us all: What are we living for; what are our motives; what guides our mood and behavior, our conduct, and our interpersonal behavior? This question would not have been put in such a sophisticated form, but it must have been implicit.

The earliest answer was derived from family experience: to please the mother and father, to please the gods, or God. Hope of reward and fear of punishment still operate strongly among us. Another answer came from tribal experience: to promote the tribal welfare even at the sacrifice of oneself. This gradually extended beyond the tribe to include the State,

---

**School of Medicine, University of California, San Francisco, California, 94122.*

and in the example set by Jesus, all humanity. Plato (427-347 B.C.) developed this ethic, it became the traditional Judeo-Christian ethic under the influence of Thomas Aquinas (1225-1274) and then Karl Marx (1818-1883) extended it to the communistic ideal of a materialistic explanation for history and economics. Meanwhile, an opposing ethical principle developed, formulated in part by Epicurus (341-270 B.C.), that we live for the purpose of getting as much personal individual pleasure out of life as possible, regardless of the society to which we may belong.

From the time of Aristotle (384-322 B.C.) to the present, many attempts have been made to develop an ethic which would effectively compromise the individualism of Stoic hedonism and the socialism of Platonic idealism. The Aristotelean injunction toward harmonious adaptation to the environment, like all subsequent efforts at compromise, has not been very successful. They all suggest the difficult *boundary conditions* in the structural hierarchy of human individuals on the one hand and their societies on the other. Bees, wasps, ants, and maybe baboons seem to have surmounted these *boundary conditions* between individuals and societies in the organizational levels of their living material. The prospect of such a development for humanity is repugnant indeed, as Aldous Huxley so bitingly showed in *Brave New World.*

THE RECOGNITION OF ORGANS

The recognition of a sub-individualar organizational level of living material seems to have come slowly. This is the appreciation of what we call *organs* – the skins, hearts, muscles, bones, livers, lungs, stomachs, brains, kidneys, and so on – the orderly associations which comprise individual *organisms*.

Organs probably began to be differentiated as a result of observations made while hunting or from observing injuries suffered in fighting, by accident, or even in sacrifice. The ancient Sumerians, for example, had a richly formulated system of prognostication associated with the character of livers observed in sacrifice and the Aztecs were certainly acquainted with hearts.

Formulation of organ knowledge grew slowly and often quite arbitrarily, as with the ancient Chinese. Once the tradition about the number of bones was established it did not change for centuries, although it was clearly wrong. The old Egyptians must have known much about bodily organs while they perfected the art of mummification of animals, birds and people. Yet, the short treatise on the heart at the near-end of the Ebers Papyrus of around 1550 B.C. is fanciful rather than informative with its four vessels to various parts of the body. These indicate recognition of stomachs, eyes, muscles, livers, lungs, spleens, and bladders.

Egyptian biomedical ideas came early into the emerging Greek world, as shown so well by Steuer and Saunders. Characteristic of Greek biomedical development was the functional hierarchy of the *four humors*. Derived from the atomism of Empedocles (490-430 B.C.), this was based on the assumed four elements of water, fire, air and earth, with their corresponding qualities of wetness, warmth, cold, and dryness. The hierarchy ranged downward from that humor thought to be most alive, blood (combining wetness with warmth, water and fire), to phlegm (combining wetness, cold, water and air), to bile (maybe urine, combining warmth with dryness, fire and earth), and to the least alive or closest to death, melancholia, or black bile (maybe the blackwater of terminal malaria, joining dryness and cold, or earth and air). It is remarkable that the adjectives describing the *temperaments* associated with excesses of any one of these humors still survive in our vernacular, in *sanguine*, *phlegmatic*, *choleric* (or *bilious*), and *melancholic*, respectively. This *humoral pathology* developed in detail by Galen (129-201 A.D.) persisted in European medicine until the 17th Century of our era.

Galen's account of the parts of the body was authoritative for centuries. It was early translated into Latin both from the structural standpoint *De anatomicis* and from the viewpoint of function *De usu partium.* These included his own sharp observations as well as those carried over from the Alexandrians, Herophilus (fl., 300 B.C.) and Erasistratus (304-250 B.C.). There was already a comprehensive anatomical background for knowledge of human bodies, with their many organs, when Mondino (1275-1326) wrote his *Anothomia,* and this in spite of the comparative neglect of the subject by the Muslim physicians and medical writers. Mondino's work was a practical dissecting manual, probably devised to aid in the legal effort at Bologna to determine causes of death. It is interesting that no significant hierarchy of anatomical organs developed. The heart, however, since Aristotle's time was thought to be paramount.

Genuine curiosity regarding the organization of our bodies seems first to have been shown by Leonardo da Vinci (1452-1519), whose scientific studies on human anatomy may have greatly helped him in becoming such a supreme graphic artist. Dissecting parts of at least ten bodies, as shown by O'Malley and Saunders, he was apparently planning a great illustrated book depicting the parts of the body and their organizational relation. Pazzini has recently tried to reconstruct this book from the great mass of notes and drawings left by Leonardo. Actually the task was first carried through, and seemingly quite independently by Andreas Vesalius at Padua in his great *Fabrica*, published so sumptuously by Oporinus of Basle in 1543. It is interesting that this year also witnessed the publication of *De revolutionibus* of Nicolaus Copernicus (1473-1543), thus achieving the Renaissance account of the structural hierarchy of both the Macrocosm, the Universe, and the Microcosm, or Humanity.

It is pertinent to note that there was no significant attempt to depict hierarchical structure among the organs of animals or of plants.[1] With the great *De motu cordis* of William Harvey (1578-1657) in 1628, there may have been some preponderance of thought in regard to the heart and circulation, but this was far from any systematic attempt to establish hierarchical values for organs. Indeed, the tendency to recognize the coordination of all the organs in maintaining the activity of the individual organism as a whole was already established. This is remarkable since there had been hierarchical structures among individuals from antiquity. Pecking orders seem to be established whenever birds or animals congregate. There are hierarchical structures of individuals within societies. In addition there were always some sort of hierarchical structures among societies.

The full recognition of organs in the structural hierarchy in organizational levels of living material came with the appearance of *De sedibus et causis morborum* in 1761 by G. B. Morgagni (1682-1771). In this important work he demonstrated the correlation between symptoms of disease in a person during life with the abnormalities found in various organs of that person at post-mortem examination. Meanwhile great technical advances came with microscopes. With their aid, Marcello Malpighi (1628-1694) was able to demonstrate the organs of plants as well as of insects, and to show their embryological development. It remained only for Marie F. X. Bichat (1771-1802) to emphasize tissues, such as blood and connective sheaths, to complete the recognition of organs and tissues as comprising an essential level of organization of living material.

---

1. In the plant world there was early recognition of the distinction between a tree and a forest to which it might belong, and the adage of not seeing the one on account of the other is an ancient one. Theophrastus (372-287 B.C.), Aristotle's pupil classified plants in a general way as Aristotle had classified animals, and seems to have recognized a sort of hierarchical system in gradation from simple to more complex, but they were both concerned chiefly with individual specimens and the groups to which those specimens belonged.

THE CELL THEORY

Although cellular structure in plants was noted by Robert Hooke (1635-1703) in his *Micrographia* of 1665, it was the influence of M. J. Schleiden (1804-1881) and Theodor Schwann (1810-1882), with their publications on the structure of plants and animals in 1838 and 1839 that established the cell theory. Cells were thought to be the units of living material. This idea was greatly supported by the recognition at this same time of protozoa and bacteria as independently existing cells, as outlined by C. G. Ehrenberg (1795-1876). The notion of cells as the basic units of living material persisted well into our own times.

Thus, in 1933, A. J. Clark (1885-1941) issued a provocative volume on *The Mode of Action of Drugs on Cells* in which he tried to relate the interaction of the units of chemicals (molecules) on the one hand, with what he still considered to be the units of living material (cells) on the other hand. In his *General Pharmacology* (Berlin: Heffter's Hndbk Pharmakolgie, 1938), he realized his error: molecules of chemicals can only react with other molecules, whether in living material or not. This may be considered to be the origin of current interest in molecular biology.

The year 1847 was remarkable for its intellectual ferment. Students were rioting against authoritarianism in Paris and Berlin; migrations of peoples westward were in full swing; practical anesthesia was achieved for surgical advance; economic depressions stirred unrest, and manifestos appeared. Karl Marx (1818-1883) and F. Engels (1820-1895) proclaimed that history and economics are explainable in materialistic terms, and that young Berlin trio, Karl Ludwig (1816-1895), Hermann von Helmholtz (1821-1894) and Emil DuBois-Reymond (1818-1896) stated that all living processes, including consciousness, are explainable in terms of physics and chemistry. These two manifestos set the pattern for the intellectual and political background of modern Russia. And in this same year, Rudolph Virchow (1821-1902) founded the famed *Archiv für pathologische Anatomie und Physiologie.* In one of its introductory essays, Virchow announced his intention of establishing pathology as *the science of disease from cells to societies.* This was the first clear statement on organizational levels of living material.

Virchow went on to develop *Die Cellularpathologie* (Berlin: Hirschwald 1858) in such a skillfully detailed manner that it remains the basis of modern and current pathology the world over. In his later years Virchow began his work on social pathology by studying anthropology. It is surprising that after more than a century we have not yet begun a systematic approach to social pathology, although its manifestations are in our midst daily.

FROM MOLECULES TO ECOLOGIES

In our own century we have extended our understanding of the organizational levels of living material at both ends of the *Virchovian spectrum* from cells to societies. With many kinds of specialized methods from high-resolution and electron microscopes to monolayer cell cultures with phase-contrast microscopy and time-lag cinemaphotography, we have come to a detailed understanding of the complexities of sub-cellular organizational units.

Among all the various kinds of cells, we have not come to agreement as to what kinds of cells constitute what kinds of hierarchical structure. The situation is similar in part to the lack of recognizing hierarchical structure among organs. All the cells which comprise an organ or tissue are considered to be necessarily coordinated for the effective functioning of the respective organ or tissue. However, there is gradual appreciation of the hierarchical primacy of germinal cells. It is from such embryological centers that all cells derive by processes of growth and differentiation which are not very well understood. Carl Ernst von Baer (1792-1876) postulated "germ layers" and this concept was extended by Wilhelm His (1831-1904) to "developmental mechanics," and further to an idea of "organizers" in cells by Hans Speman (1869-1941).

Attention became focused on the nuclei of cells, and their make-up. Hugo de Vries (1848-1935) in resurrecting the heredity principles discovered by Gregor Mendel (1822-1884) emphasized the importance of chromosomes, and Thomas Hunt Morgan (1866-1943) was able to map the location of some of the Mendelian factors on specific spots on chromosomes, thus developing *The Theory of the Gene* (New Haven, Conn: Yale University Press). His great work was mostly done on fruit-flies.

The dynamic character of cellular components has been brilliantly shown by Charles M. Pomerat (1905-1964) in his superb cinematographic studies of monocellular layers under phase-contrast microscopy, with a minimum of external interference. One can excitedly follow the rotation of nuclei within their sacs, with extrusion of lipo-protein granules; the formation and fading of vacuoles; the activity of the Golgi apparatus; and the amazing activity of mitochondria. Specific membranes seem to characterize nerve, muscle and blood cells. Many other cells seem to have phase boundaries, permitting pinocytosis or cell drinking, as described by Warren H. Lewis (1870-1964). Centrosomes had been described by Walther Flemming (1843-1905) in 1877, but the detailed mechanism of cell division is still being studied. The dynamics and the structures involved again can be well shown by phase-contrast microscopy and cinematography. It is clear the cells are vastly complex and highly organized structures with a rich assemblage of co-ordinated sub-cellular components.

Inevitably the molecular components of sub-cellular structures have been investigated. Genetic material on chromosomes has been brilliantly analysed, with extensive popularization of the dramatic exposition of the double-helical structure of desoxyribose-nucleic acid, the complex protein carrying the genetic coding mechanism for all living material. This was accomplished just a few years ago by Francis H. C. Crick (1916-), James Dewey Watson (1928-), and M. H. F. Wilkins (1916-). Watson was the brash American who gave the all-too-human side of the scientific story in the gossipy book, *The Double Helix* (New York: Atheneum 1968), published last year.

All the sub-cellular units of living material are now considered to be molecular complexes of various kinds, ever in dynamic activity maintaining a precarious "internal steady state" as suggested by Claude Bernard (1813-1878), or "homeostasis" as outlined by Walter Cannon (1871-1945). This concept runs through the whole gamut of organizational levels of living material from molecules to ecologies.

The molecular equilibrium in living material goes from inorganic ions (especially sodium, potassium, calcium, magnesium, chloride, sulfate and phosphate) to organic ions; from amines, steroids, and amino acids to proteins, fats, carbohydrates (with their many metabolic products); to enzymes and co-enzymes, with astonishing detail on the various cycles of interaction which these various chemicals can undergo. Some large molecules of desoxyribose-nucleic acid (DNA), often with protein coatings, seem to exist "free" as viruses, and may even be crystallized. What are they then, alive or not? They can metabolize and reproduce, even though they may be crystallized. At this point the organization of living material fades into the non-living, and the distinction becomes irrelevant.

Even while scientific exploration into the details of sub-cellular and molecular levels of organization of living material was attracting the major attention of biomedical scientists, Rachel Louise Carson (1907-1964), a gentle but keen and determined woman, sharply focused popular interest at the other far end of the range of organizational levels of living material, by emphasizing the supreme significance of ecologies. Her *Silent Spring* (Greenwich, Conn.: Fawcett-World Library 1964) made a profound impression, inspiring conservationists all over the world, and arousing wide-spread alarm that we may be utterly ruining our environment by the chemical compounds we spew forth, whether in the pollution of airs and waters, or in the deliberate use of pesticides and herbicides.

Biologists had long tried to explain the importance of trying to preserve the "balance of nature." Justus von Liebig (1803-1873) had long ago demonstrated the grand cycle of carbon, nitrogen, and oxygen through plants and animals and their environments in keeping a dynamic equilibrium throughout the living world. His concept suggested the domino effect of molecular interactions at a basic organizational level of living material sweeping through cells, organs, individuals, societies and into environments.

IN PROSPECT

Although there is empirical appreciation of the procession of events in living material from such macromolecules as DNA to such ecologies as our badly abused planet, there is little understanding of the way by which the hierarchical structure operates. It seems to be statistical, involving huge numbers of particulars. These numbers, however, progressively diminish as one goes from molecules to ecologies. Molecules of insecticides such as parathion (o,o-diethyl-p-nitrophenyl phosphorothiate) for example, react with enzyme molecules in mitochondria of insect cells, blocking the enzyme destruction of acetylcholine which helps transmit nerve impulses. As acetylcholine molecules accumulate, subcellular functions are disturbed, so that cellular activity is impaired, so that organs no longer operate, so that the co-ordinated individual insect dies, so that all the similar insects in the area die, so birds no longer have an adequate food supply, so that the ecology is disrupted. Or the ecological milieu may be destroyed, as in the case of Lakes Erie and Constance, by pollutants which permit such a growth of algae that the oxygen supply is exhausted and all life in the lake, including fishes, that depend on oxygen, dies. The process goes from molecules to ecologies, and yet *boundary conditions* remain between molecules, sub-cellular units, cells, organs, individuals, societies, and ecologies.

It is interesting that we are unable to predict with certainty as we move upward in this hierarchical structure. No matter how much we know about brains, hearts, lungs, livers, skins, blood and so on at an organ and tissue level, can we say with any surety what any particular individual mammal will be like. No matter how much we may know about individual people we cannot tell how they may act when they form a mob. Nor can we surmount the boundary conditions any better when we move downward in the biological hierarchical structure. No matter how much we may know of a human individual we cannot say surely what that person's organs and tissues may be like until we examine them.

It seems then that our knowledge of the steps in the biological hierarchy is statistically determined. We may be able to talk in general about any one particular organizational level, but in order to speak with surety about any one particular item in the hierarchy, whether DNA molecule, subcellular unit such as a particular membrane, some single cell, some one organ, some single person, some particular society, or some particular ecological unit, we have to examine that particular item itself.

Erwin Schrödinger (1887-1961) in his brilliant little volume, *What is Life?* (Cambridge: Cambridge University Press. 1944) indicates the amazing events which can occur at a molecular level in genetic material in relation to energy quanta. Here finally we seem to be approaching an understanding of the origin of species, that phrase which Charles Darwin (1809-1882) used improperly in 1859, when he showed how species survive through adaptation to changing environments. As R. N. Giere indicates, in criticizing Michael Polanyi's article on "Life's Irreducible Structure" (*Science*, 160: 1308-12):

> While it may be of some interest to think of the universe as a hierarchy of systems, each providing boundary conditions for 'lower' systems, it has not been shown that any but physico-chemical laws are needed throughout the hierarchy.
> (*Science*, 162: 410)

Yet, we have mainly faith to hold to the 1847 manifesto of Ludwig, Helmholtz and DuBois-Reymond that all living processes, including consciousness, are explainable in terms of physics and chemistry.

The matter is well put by our colleague, Cyril Stanley Smith:

> Science now relates to the two extremes of elementary atomistic physical chemistry on one hand and averaging thermodynamics on the other. But why cannot science develop a new approach encompassing the whole range? I am not as pessimistic as Polanyi, for I see in the complex structure of any material–biological or geological, natural or artificial–a record of its history, a history of many individual events each of which did predictably follow physical principles. Nothing containing more than a few parts appears fully panoplied, but it grows. And as it grows, the advancing interface leaves behind a pattern of structural perfection or imperfection which is both a record of historical events and a framework within which future ones must occur. Deoxyribosenucleic acid is simply a mechanism to save time in reaching higher levels of organization, though, of course, with severe limitation of possible structures . . . Is there not possible an intermediary science using the structure that exists . . . both as a key to history and as a framework for continuing process?
> (*Sciences*, 162: 637-644)

George Sarton, the great historian of science, in whose honor Doctor Smith was speaking at the 1967 New York meeting of the American Association for the Advancement of Science, would have been in full accord.

Evolutionary change in living material is usually considered to be due to the operation of external environmental factors. But Lancelot Whyte has well stressed the importance of internal factors built into living material which may operate quite as significantly as external factors in guiding the development of life's organizational levels. It is an understanding of these internal factors in the hierarchical structure of life that we must seek.

■

Physical Conditions for Primitive Functional Hierarchies

Howard Pattee*

The origin of life problem is the context in which I began thinking about hierarchies. The origin of life is perhaps the most mysterious hierarchical interface of all, but at the same time I believe it may present one of the most instructive approaches to general hierarchical control problems. This is because the lower level pre-life processes are ordinary physics and presumably subject only to precise laws which do not include extra hierarchical rules or constraints. However, to be recognized as "alive" a collection of matter must exhibit some additional integrative function by exerting a collective control over the individual molecules. This integrative function is what characterizes hierarchical control.

Hierarchical systems raise two types of questions. Viewed from the upper level of the hierarchy the existing *constraints* are taken for granted and the significant question seems to be *How does it work?* The answer found from this perspective always amounts to the discovery that the parts obey the laws of the lower level. Thus we find that if we take apart a working machine, like a watch, there is no detail of motion which evades physical laws. People with this perspective often claim that molecular biologists have reduced life to physics since they have taken the cell apart like a watch and found that no detail evades physical laws [1].

On the other hand, viewed from the lower level of the hierarchy it is the laws of motion themselves which are assumed to be inexorable, and the significant question seems to be *How could the constraints arise?* The answer usually given from this

---

**Hansen Laboratories of Physics, Stanford University, Stanford, California 94305.*

1. All the authors on molecular biology I have read tacitly assume that the classical idea of a deterministic machine is a good physical analogy to living matter. e.g., D. E. Wooldridge, *The Machinery of Life,* (New York: McGraw Hill, 1966). F. Crick, *Of Molecules and Men,* (Seattle: University of Washington Press, 1966). No one, except Polanyi (1968), points out that machines are designed and built only by man, and are therefore a biological rather than a physical analogy.

perspective amounts to the conclusion that the constraints are not clearly derivable from the laws of the lower level (Polanyi 1968). To this extent reduction appears very nearly impossible, and this is why some of us find the existence of hierarchies such a mystery.

If anyone takes care to formulate both of these questions with more logical precision, then I think he will find that both answers are correct for their respective questions. I believe most of us here feel that the second question is far more significant than the first. Nevertheless, either of these answers alone has tended, for hundreds of years, to stimulate great disputes. Since these questions arise from disjoint perspectives, the arguments are often largely polemical. Of course we do not desire or expect to avoid arguments over this discussion. All I can say is that I am not at all satisfied with the claims of either side, that physics explains how life's constraints work or the claim that physics cannot explain how life's constraints arose.

THE CONCEPT OF HIERARCHY

To begin I shall limit my use of the idea of hierarchy to *autonomous* hierarchies; that is, to collections of elements which are responsible for producing their own rules, as contrasted with collections which are designed to have hierarchical behavior by an external authority. I want to talk only about what might be called natural hierarchies rather than artificial or supernatural hierarchies, such as man-made machines or "special creations" of any kind. Secondly, I shall assume that all my examples are a part of the physical world and that *all the elements obey the normal laws of physics.* This does not mean that I assume a reductionist attitude. The question of what reduction can mean will become clearer, I believe, only after we discover the necessary physical conditions for a hierarchical interface. Thirdly I shall limit my definition of hierarchical control to those rules or *constraints* which arise within a *collection* of elements, but which affect *individual* elements of the collection. This is the normal biological case where, for example, in society a set of laws is enacted by the collective action of the group but applied to individuals of the group; or in the development of the organism, the collective interactions of neighboring cells control the growth or genetic expression of an individual cell; or in the enzyme where collective interactions of many bonds control the reaction of an individual bond.

Finally we must recognize the essential characteristic of hierarchical organization that the collective constraints which affect the individual elements always appear to produce some *integrated function* of the collection. In other words, out of the innumerable collective interactions of subunits which constrain the motions of individual subunits, we recognize only those in which we see some coherent activity. In common language we would say that hierarchical constraints produce specific actions or are designed for some purpose.

Right here I shall stop my description rather abruptly, since in talking about "function" I have passed over the hierarchical interface which always causes so much argument. Let me return instead to the first three conditions for a hierarchy: (i) autonomy, that is, a closed physical system, (ii) elements in the system which obey laws of physics, and (iii) collections of elements which constrain individual elements. I want first to express these conditions in the language of mechanics so that we can see the implications of these conditions as simply as possible.

STRUCTURAL HIERARCHIES

Descriptions of nature using the language of physics usually satisfy our first condition of autonomy by assuming a closed system. In classical mechanics the elements or particles in this system are said to have a certain number of degrees of freedom, which is just the number of variables necessary to describe or predict what is going on. Our second condition is that the particles of the system follow the laws of motion. Classically this means that given their initial positions and velocities at a given time, the trajectories of the particles can be predicted in the future or explained in the past with arbitrarily high precision. But if we are restricted to classical physics, there is no way in which the third condition can be satisfied because it requires a "collection" of particles which constrains individual particles. The implication here is that some particles join together in a more or less permanent collection; otherwise the "collection" would only be transient and would depend crucially on the initial conditions. It was one of the serious difficulties of classical physics that there was no inherent dynamical reason why collections should ever form permanently. In quantum mechanics, however, the concept of particle is changed, and the fundamental idea of a continuous wave description of motion produces the "stationary state" or a local time-independent "collection" of atoms and molecules. Since these local collections are constantly being perturbed, they are not really permanent, but have lifetimes which increase with the energy of the interactions which hold them together, and decrease with the thermal energy which knocks them apart. Although there are several types of bonds between atoms and molecules we need to distinguish only two – the strong and the weak bonds. The structures held together by the strong chemical bonds will have lifetimes much longer than structures held by weak bonds.

So far our simple physical description is useful up to the level of polymers and crystals, but now we need to see how such collections can "constrain" *individual* monomers or atoms which make up these collections. Up to this point, our description of matter is "normal physics" at the level of atoms and molecules, but the concept of "constraint" begins to sound like we are introducing new rules. What is the physical meaning of a constraint? The concept of "equation of constraint" was in fact first necessary in classical physics because of the lack of any dynamical process to explain the permanent loss of degrees of freedom of collections of matter in solid bodies. Another type of constraint is the boundary condition which limits the values of certain degrees of freedom independently of the equations of motion – e.g., when a particle is confined by a box. Both solid bodies as well as walls of boxes could be considered as collections of particles which influence the motion of individual particles, and so they fulfill the second condition of our definition of hierarchy. But while we know that solids can form spontaneously from individual particles, constraints such as boxes are usually designed by experimenters with some "higher" purpose in mind, and in this case our first condition of autonomy would not be satisfied. However, it is primarily the stationary-state solutions of the quantum mechanical equations of motion which account for permanent constraints.

From such apparently simple beginnings we can see the origin of what are often called *structural* hierarchies. The richness as well as the orderliness in all the natural patterns of collections of molecules and crystals could be described as a selective and more or less permanent loss of degrees of freedom among many elements. Many scientists and philosophers will assert on principle that such hierarchical structure is entirely reducible to quantum mechanics. As is often the case, those experts who actually study the details are seldom so easily convinced. For example, Cyril Smith (1968) has pointed out that new levels of structural hierarchies usually depend on the appearance of an imperfection in the old level. But what do we mean by an "imperfection"? Which imperfections lead to new levels of organization, and which lead to greater disorder?

FUNCTIONAL HIERARCHIES

In spite of the enormous complexity which we can find in structural hierarchies, there is still something missing. There is seldom any doubt that such structures are lifeless. What is missing is some recognizable "function". No matter how intricate a structure may be, permanence is not compatible with the concept of function. Function is a process in time, and for living systems the appearance of time-dependent function is the essential characteristic of hierarchical organization. To achieve function by permanently removing degrees of freedom in a collection of elements would be impossible. Instead the collection must impose *variable* constraints on the motion of individual elements. In physical language these amount to *time-dependent boundary conditions* on selected degrees of freedom. Furthermore, the time dependence is not imposed by an outside agent, but is inseparable from the dynamics of the system. Such constraints are generally called non-holonomic (non-integrable), and have an effect which is like modifying the laws of motion themselves. For example, the enzyme is not just a permanent linear string of amino acids residues, nor a permanently folded three-dimensional molecule. An enzyme is a time-dependent boundary condition for the substrate, which through the collective interaction of many degrees of freedom controls a few degrees of freedom so as to speed up the formation of a strong bond. Nor is it the essential peculiarity of the enzyme that it is a very *complicated* dynamical system. Any system with that many degrees of freedom is dynamically complicated. What is exceptional about the enzyme, and what creates its hierarchical significance, is the *simplicity of its collective function* which results from this detailed complexity.

To put the problem of dynamical hierarchical control in a more general way, it is easy to understand how a simple change in a single variable can result in very complicated changes in a large system of particles. This is the normal physical situation. It is not easy to explain how complicated changes in a large system of particles can repeatedly result in a simple change in a single variable. It is this latter result which we interpret as the "integrated behavior" or the "function" of a hierarchical organization. Thus, we find *structural* hierarchies in all nature, both living and lifeless, but we see *functional* hierarchies as the essential characteristic of life, from the enzyme molecule to the brain and its creations.

However our recognition of *function* as having to do with a simple result produced by a complicated dynamical process is not useful unless we can give some physical meaning to the idea of simplicity. The problem is that the concept of simplification is not usually associated with the physical world, but rather with the observer's symbolic representations of this world. The world is the way it is. Only an observer can simplify it. In fact it is the assumption that the elementary motions are complete and deterministic that makes the generation of hierarchical rules appear so difficult. The hierarchical rule is superimposed upon a lawful system which is already completely deterministic. How can this be done without contradiction?

As far as I can see, this has never been done in physics without introducing what amounts to a measuring device or an observer. Unfortunately, since measuring devices and observers are usually associated with the brain, this does not resolve the contradiction, but only substitutes a human language hierarchy which is a harder problem than the one we are asking. I want to think of the most elementary configurations of molecules in which we recognize some simple objective function. So again the question arises: How can a lawful system of atoms which is maximally deterministic superimpose an additional functional rule or constraint upon its detailed motions?

And again, the only answer must be that *the concept of functional constraint implies an alternative way of representing the detailed motions.* But in a closed physical system there is no observer to represent the system in a different way. Therefore we are left with the idea that if we can recognize a simple hierarchical function in an isolated dynamical system, then we should also be able to recognize an internal *representation* or *record* of the system's own dynamics. Autonomous hierarchical function implies some form of self-representation. In other words, we may partially resolve the appearance of hierarchical order on an already completely ordered set of elements by saying that hierarchical rules do not apply to the elementary motions themselves but to a record of these motions. Before we look at some examples of simple molecular collections which may exhibit internal records, let us see under what conditions our own hierarchic representations of physical systems arise.

DESCRIPTIVE HIERARCHIES

The hierarchical levels of our languages contain some of the deepest mysteries of logic as well as epistemology, but I believe they also contain a clue to the physical problem of the hierarchical interface. We have already mentioned the crucial interface between the strictly causal language of dynamics and the probabilistic language of statistical mechanics which has produced much distinguished controversy. I shall try to avoid the intricacies of the general arguments by using a simple example as an illustration.

When we speak of the elementary laws of mechanics we mean the laws that describe as precisely as possible how each degree of freedom changes in time, given the initial conditions and boundary conditions. These equations of motion are universal and apply to all detailed motions which take place in the system. In one sense, therefore, all additional information about the system is either redundant or contradictory. But if we are trying to describe, say, 10^{23} molecules in a box, it is obvious that measuring or following each degree of freedom is impossible. However, as *outside observers* we have learned to recognize and define collective properties of molecules, such as temperature and pressure, which allow simple and useful measurements on the gas in the box. It is significant that these properties were measured long before their "molecular basis" was known, just as many hierarchical biological functions were accurately described before a "molecular basis" was discovered. In physics it was the later discovery of the molecular dynamics which began the controversial attempts to reduce thermo-dynamical description to mechanical description by rigorous mathematical arguments. Perhaps these attempts can be characterized as very nearly successful – but not quite. This result is not trivial, since "not quite proved" in mathematics is like "not quite pregnant" in biology.

We may look at the problem as arising from the inability of the formal mathematics to predict what collective properties of complicated systems will produce simple, significant effects in the physical world of the observer. In other words, while there is no question that the detailed equations of dynamics can be used to calculate previously well-defined averages or collective properties, there is no way to predict from only the dynamical laws of the system which definitions of collective properties are significant in terms of what we actually can measure. Thus in one sense we can derive the pressure in terms of a suitable average of dynamical variables, if we are given a precise definition of pressure; but this definition of pressure is not determined by the equations of dynamics. The concept of pressure appears useful only when the dynamical system is embedded in a particular type of observational environment.

More generally we may say that a physical system which appears complete and deterministic with the most detailed symbolic representation, can appear incomplete and probabilistic only with a new representation which relinquishes some of the detail. The new representation must therefore come about through the *combination* or *classification* of the degrees of freedom at the most detailed level so as to result in fewer variables at the new level. Formal reductionism fails simply because the number of possible combinations or classifications is generally immensely larger than the number of degrees of freedom. What must always be added to define a new representation is the rule of combination or classification which tells us how to simplify the details. In statistical mechanics this rule is usually a hypothesis of randomness or ergodicity, but the ultimate justification for any such rule is that it results in a more useful description of the system in the observational environment in which the system is embedded.

What can it mean, then, for a collection of particles to form an *internal* simplification or *self*-represenation? What is the meaning of an "observational environment" for a system which is closed? Clearly in an autonomous hierarchy there must be an internal separation of some degrees of freedom from other degrees of freedom which become constrained to impose collective and time-dependent boundary conditions on individual degrees of freedom. While we know such integrated systems exist in cells, and can design machines which operate in this way, we are still baffled by the spontaneous origin of this type of constraint.

It is, in fact, a characteristic difficulty of hierarchical interfaces in biological organizations that their actual operation may appear quite clear while their origin is totally mysterious. The genetic code is a good example of a crucial hierarchical interface that is clear in its operation, but mysterious in its origin. One might wonder, in fact, if there is some inherent reason why a hierarchical organization obscures its own origins. Since it is one general function of hierarchies to simplify a complex situation, Simon[2] has suggested that if there are ". . . important systems in the world that are complex without being hierarchic, they may to a considerable extent escape our observation and understanding." Putting it the other way around, I would also suggest that "being hierarchic" requires that the system control its dynamics through an internal record, which has some aspects of "self-observation."

THE LOWEST HIERARCHY

But this is only evading the question. Let us see if we can clarify the problem of hierarchical origins by looking at collections of molecules of gradually increasing complexity, watching closely for any signs of internal *classification* or *recording* processes which are the essential conditions for a

2. In addition to emphasizing the essential correlation between state and process languages in any functional hierarchy, Simon (1962) characterizes hierarchical organizations as "nearly decomposable" by which he means that the state space is larger than the trajectory space. This is nearly equivalent to what I call a non-holonomic constraint.

simplification of the detailed dynamics. If we can imagine such collections, then we may go on to ask if this internal simplification is inherently self-perpetuating, or if there appear to be additional conditions which must be satisfied to establish a persistent hierarchical organization of molecules.

Perhaps the simplest interesting level of complexity is crystal growth. First, consider an ideal, ionic crystal growing in solution. One might try to apply our hierarchical conditions by saying that the crystal surface, with its alternating positive and negative sites, "classifies" the incoming ions, and by permanently binding each ion to a site with the opposite charge forms a "record" of the classification interaction. Now while this may be grammatically correct, it is really only a redundant statement. There is no real distinction here between the physical interaction of the ion and the binding site and what we have called the "classification" and "record" of this interaction. They are all the same thing. Furthermore, each ion's interaction is local and direct and does not involve the dynamics of any large collection of ions or any delay. Therefore, although we may call this ideal crystal an example of hierarchical structure, I would not say that it exhibits hierarchical control over its dynamics.

Let us go on, then, to a more realistic level. Consider crystal growth which is produced by an imperfection, such as a screw dislocation. This is a *statistical* process which requires more than one atom or molecule to be in metastable positions. In time these atoms would shift to stable positions if there were no further growth. But this screw-dislocation structure increases the rate of growth by many orders of magnitude, all the time maintaining its special structure even though the original collection which first introduced the dislocation has been buried deep within the crystal. In this example, I believe a much stronger case can be made in favor of calling this a kind of hierarchical control. First, the constraint which controls the growth dynamics is not simply the direct interaction between local atoms, but involves the *collection* of atoms which makes up the dislocation. Second, this collection is not the original dislocation, but a *record* of a dislocation which is propagated over time intervals which are very long compared to the rate of addition of the individual atoms. However it is difficult to distinguish a classification process in this example since all the atoms are identical.

As a third more complicated example, then, imagine a protoenzyme made up of only two types of monomers in a linear chain. Suppose this particular sequence of monomers folds up into a catalyst which speeds up the polymerization of only one type of monomer. For this specific catalytic reaction to occur we must express the fact that the folded polymer can distinguish one type of monomer from the other, and on the basis of this distinction alter the dynamics of each correct type of monomer so that it reacts much faster. Or in other words, we may say that this sequence of monomers *classifies* its elements and *records* this classification by forming a single, permanent bond between monomers. Now is there anything wrong with calling this process a form of hierarchical control?

In so far as the polymer sequences are no longer determined directly by the dynamical laws of the individual monomers (including their inherent reactivities), but by the constraints of a special polymer which speeds up the formation of a particular sequence, this might be called hierarchical dynamics. But now I think we have some problems of autonomy. First, this specific catalyst was invented by me, and although we know such specific catalysts do exist as enzymes, my invention simply evades the origin problem, as well as the physical problem of how such specific catalysts work. However, I have in mind a problem which is much more important. I think this example misses the essence of hierarchical *control.* We may indeed have in the catalyzed homopolymer a kind of simple record of a rather complex dynamical interaction, but the record has no further effect.

The trouble is that in the context of autonomous hierarchies, what constitutes a "record" must be indicated within the closed system itself and not by what I, as an outside observer, recognize as a "record." Obviously to generate autonomous hierarchical control the record must be *read out* inside the system. The time-independent constraints formed by the permanent strong bonds must in turn constrain the remaining degrees of freedom in some significant way. This was the case in the previous example of screw-dislocation crystal growth where the dislocation structure was both a record of a past collective imperfection and a catalyst for the future binding of individual atoms. Cyril Smith (1968) sees this process as requiring a new description somewhere in between the detailed dynamics of atoms and the simple, stationary averages of thermodynamics. He sees all complex structure as both a record and a framework: "... the advancing interface leaves behind a pattern of structural perfection or imperfection which is both a record of historical events and a framework within which future ones must occur."

Returning to the copolymer system, we see that it may indeed fulfill the function of a record of past events, but the homopolymer record which was catalyzed does not act as a framework for future events. To provide autonomous hierarchical control, the catalyzed product of one copolymer must lead to the catalysis of other specific reactions. Furthermore, if the record is not to be lost, each catalyzed sequence must in turn catalyze another, and so on indefinitely. Now clearly such a sequential process can be divergent or convergent depending on the rules of specificity for the catalyses. Even if we assume that there is no error in these rules, a divergent record would never be recognized. One might say, in this case, that the system's self-representation is as complex as the system itself. But I think no *under*constrained system would produce such a chain of catalysts. The starting record would simply disintegrate.

Going back now to the hierarchical control in the screw-dislocation crystal growth, we may look at this example as the other extreme. Here the classification and record possibilities are trivially *over*constrained. Since there is only one distinguishable type of monomer, there can be no classification and hence no linear record. The "record" is not distinguishable from the three-dimensional structure which is also the functional catalytic site. The same problem of *over*constraint could, of course, occur in a copolymer system where, say, an alternating-sequence polymer acts as a tactic catalyst for the same alternating sequence. But this is the point of these examples. I want to show that even the simplest hierarchical organization requires a balance between the number of degrees of freedom of its elements, the number of fixed constraints which function as a record, and the number of flexible constraints which encode or transcribe the record.

Of course from this simplest conceivable level of molecular assembly which exhibits a potential *classification-record-control* process, we should not expect to find the nature of hierarchical interfaces at all levels. Even these simple examples present unanswered questions. But in following the necessary physical steps leading from the dynamics of individual units to the collective control of individual units, I believe we can gain some insight into the spontaneous generation of hierarchical organization.

First, we see that the individual particles or units follow more or less deterministic laws of motion. These units were atoms or molecules in my examples, but we may also think of the units as cells, multicellular individuals, or population units. The "motions" of these larger units are not as deterministic as the motion of atoms, but they have definite patterns of unit behavior. Second, there are forces between units which produce constraints on the individuals. These forces cause permanent aggregations of units which act as relatively fixed boundary conditions on the remaining individuals. By "relatively fixed" I mean that the rate of growth or change of these aggregations is slow compared to the detailed motions of individual units. These strong forces form what we called structural hierarchies, but they are essentially passive constraints.

The third stage is crucial and, as we might expect, the most mysterious. If the fixed constraints are not too numerous, that is, if the aggregations are not too rigid, then *weak forces* become important in the internal dynamics of the aggregations and through this *collective* dynamics the aggregations can form *time-dependent boundary conditions* for the other individual units. This type of flexible or non-holonomic constraint reduces the number of possible trajectories of individual units without reducing the number of degrees of freedom. This amounts to a *classification of alternatives* which leads us to now use the higher language of information or control. The specific catalyst or enzyme is the simplest example of such a dynamical constraint; but at any level of hierarchical control where there are ordinary molecules which also act as messages, or where simple physical objects are said to convey information, there must be the equivalent of such dynamical constraints which classify alternative motions by leaving a record of their collective dynamical interactions.

As we said earlier, it is in the simplicity or relevance of these records or messages that we recognize hierarchical control; but how this simplicity originates remains a mystery. In practice, when a dynamically complex system exhibits simple outputs or records of its internal motions we switch languages from the detailed dynamical description to a higher language, which relinquishes details and speaks only of the records themselves. We might think of our simplified language as an *effect* necessitated by a system that is too complicated to follow in detail, as in the case of our thermodynamic description of a gas. On the other hand, in systems which exhibit autonomous hierarchical organization, it is the internal collective simplifications which are the *cause* of the organization itself. In this sense, then, a new hierarchical level is created by a new hierarchical language. Simon (1962) has come to a similar conclusion from observing a broad class of hierarchical organizations. He calls the lower level language a detailed "state description" and the upper level language is simple "process description." But the fact remains that whether it is the system-observer interface in physics, the structure-function interfaces in biology, or the matter-record interface in the most primitive molecular hierarchies, these levels are presently established only at the cost of creating separate languages for each level.

CONCLUSION

I have described the simplest examples I can imagine of what might be called incipient molecular hierarchies. I have used only a rough, semi-classical language, and have not even touched on the crucial question of how specific catalysis or classification processes could be described in the deeper quantum mechanical language.[3] Nevertheless, I find the physical concreteness of these simple examples very helpful in sorting out which conditions are most essential for establishing a hierarchical interface.

What we find is that even the lowest interesting example of a hierarchical interface is beset with precisely those difficulties that we find in all hierarchical structures, namely, that each side of the interface requires a special language. The lower level language is necessary to give what we might call the legal details, but the upper level language is needed to classify what is significant. As Polanyi (1968) has so clearly pointed out, living organizations are not distinguished from inanimate matter because they follow laws of physics and chemistry, but because they follow the constraints of these internal, hierarchical languages.

It is therefore difficult for me to escape the conclusion that to understand even the simplest biological hierarchies, we will have to understand what we mean by a record or a language in terms of a lower level language, or ultimately in terms of elementary physical concepts. Physicists have worried about the inverse problem for many years. In fact a large part of what is

---

3. These questions are discussed elsewhere (Pattee 1967, 1969).

called theoretical physics is a study of formal languages, searching for clear and consistent interpretations of experimental observations. Biologists have never paid this much attention to language, and even today most molecular biologists believe that the "facts speak for themselves." Hopefully, as these facts collect, biologists, too, will seek some general interpretations. All these facts tell us at present is that life is distinguished from inanimate matter by exceptional dynamical constraints or controls which have no clear physical explanation. We will not find such an explanation by inventing new words for *our* description of each level of hierarchical control. Instead, we will have to learn how collections of matter produce *their* own internal descriptions.

ACKNOWLEDGEMENT

This study is supported by the Office of Naval Research Contract Nonr 225(90).

REFERENCES

Pattee, H. H. 1967. "Quantum Mechanics, Heredity, and the Origin of Life." *J. Theoret. Biol.* 17:410-420.

——. 1969. "The Problem of Biological Hierarchy." In *Towards a Theoretical Biology,* Vol. III., ed. C. H. Waddington. Edinburgh: Edinburgh University Press.

Polanyi, M. 1968. "Life's Irreducible Structure." *Science* 160:1308-12.

Simon, H. A. 1962. "The Architecture of Complexity." *Proc. Amer. Philos. Soc.* 106:467-82.

Smith, C. S. 1968. "Matter Versus Materials: A Historical View." *Science* 162:637-44.

■

Hierarchical Organization in Automata Theoretic Models of Biological Systems

Robert Rosen*

Most, if not all, of the systems we are familiar with which exhibit a strongly hierarchical character are biological in origin. Moreover, it is a striking fact that within biology, the various kinds of hierarchical systems which arise are remarkably similar to one another in overall organizational structure. Here, we shall be concerned with a number of general problems which arise in the modelling of hierarchical systems, illustrated in two important concrete situations which exhibit functional analogies most clearly: cellular biology and the biology of the central nervous system.

I would like to begin, with some effrontery, by proposing a simple operational definition of a hierarchically organized system, and illustrate how this definition applies to the two special cases of interest. Next, I undertake to briefly discuss the one situation in the sciences in which the problem of hierarchical organization has been at all successfully treated in order to draw some general criteria which must be satisfied by any comprehensive theory of hierarchical systems. Finally, we investigate to what extent these principles are satisfied within the confines of one kind of theoretical framework which has been proposed for the description of biological processes, namely, the theory of automata. Naturally, in the course of a brief discussion we can only scratch the surface, but it is hoped that some of the important relevant ideas have been identified, and some valid preliminary conclusions have been drawn.

HIERARCHICAL ORGANIZATION

At the outset, then, I would like to claim that *a hierarchically organized system is simply one which is (a) engaged simultaneously in a variety of distinguishable activities, for which we wish to account, and (b) such that different kinds of system specification or description are appropriate to the study*

**Center for Theoretical Biology, State University of New York at Buffalo, Amherst, New York, 14226.*

of these several activities. In my view it is property (b) which is decisive for a hierarchical organization; a system may be doing many things simultaneously, but the idea of a hierarchical organization simply does not arise if the same kind of system description is appropriate for all of them.

The essential problem in studying a hierarchically organized system in the above sense, is to specify the relationships which exist between the system descriptions appropriate to the several levels in the hierarchy. We would especially like this specification to be predictive; i.e., to be able to use information pertaining to any particular level of the hierarchy to obtain information about the activity of the system at other levels. We are most accustomed to try to pass *downward* in the hierarchy, since as Bradley (1968) has succinctly pointed out, what we call a *model* for one level of system activity is generally a system description at the next level down. We also have one case in which it is possible to pass *upward* in a hierarchy; this is statistical mechanics, which we discuss in detail below.

One important special case of the general problem of hierarchically organized systems is: to what extent does a knowledge of the *lowest* relevant level of system activity determine its properties at higher levels? This is the essential problem of *reductionism* in biology, seen most clearly in the claims of molecular biology. Their assertion is that the lowest relevant system description is at the biochemical (or even quantum-theoretic) level, that such a description is effectively attainable and effectively determines all higher-level descriptions. I have discussed elsewhere (Rosen 1968) the problem of obtaining system descriptions at the biochemical level, while other authors (Elsasser 1966, Polanyi 1968) have concerned themselves with the possibility of effectively drawing inferences from such a description, even assuming that one could be forthcoming. More of this later.

Nevertheless, what the reductionists are attempting to do is to establish a kind of *anchor level* in the hierarchy; some specific distinguished level which is somehow more significant and convenient than the others. This is a natural thing to do; most of our intuitions concerning systems have developed from our experience of non-biological systems drawn either from physics or the world of engineering artifacts which we construct for ourselves, and which, being at best feebly hierarchical, do possess such a distinguished level. Previously in biology this distinguished role fell to the *cell*; the cellular level, the basic level of biological structure and function from which we could pass upward to tissues, organs, whole organism physiology, and even ecology, and conversely pass downward to subcellular levels; organelles, macromolecular complexes and biochemistry. The main significance of the cell theory, in fact, was that it offered us an anchor in the biological hierarchy which stood about midway in the hierarchy and from which we could pass up or down; molecular-biological reductionism offers us another anchor, but one from which we can only pass upward.

The cell theory of organization is still very strong in our views about the organization of the central nervous system, except that we use the word *neuron* instead of cell. Indeed, the central nervous system is often described in terms appropriate to an independent organism with the neural level as the anchor from which we can pass upward to problems of memory, perception, learning, etc., to psycho-physics, organized behavior and psychology, and downward to synaptology, excitation and other phenomena.

SYSTEM DESCRIPTION IN BIOLOGICAL SYSTEMS

There are two basic kinds of system description which have been used at the neural level; continuous-time descriptions and discrete-time descriptions. The continuous-time descriptions typically take the form of systems of differential equations, while the discrete-time descriptions belong to the theory of automata. (We might note parenthetically that for our purposes it is not necessary for us to be concerned with comparing and evaluating individual neural descriptions, of either discrete or continuous type; our interest here is in passing between organizational levels, and will be valid regardless of the particular neural description adopted, the choice of such a particular description being a technical problem entirely within a single level.) In the continuous-time descriptions, we chose a set of state variables for the neuron, say $x^1, \ldots, x^n$ whose instantaneous values decide whether or not the neuron is excited at that instant, and which satisfy a set of equations of the form

$$dx_i/dt = f_i(x_1, \ldots, x_n) \qquad (1)$$

Any state variable which augments excitation is called *excitatory*; one which depresses excitation is called *inhibitory*. It was early recognized (Rashevsky 1938) that such neurons could be strung into *networks* similarly described by a system of equations of the form (1), and that with a little ingenuity networks could be constructed which would possess functional properties reminiscent of behavioral properties of organisms; e.g., that could "learn" and "discriminate".

The same underlying idea, though in purely digital form, was developed by McCulloch and Pitts (1943) who showed how discrete-time "formal neurons" could be constructed which shared with real neurons the basic functional properties of excitation, inhibition, threshold, temporal summation, etc. They showed that one could build connected arrays of such formal neurons (neural networks) which could realize *any* behavior that one could write down, including presumably any behavior the brain itself might be capable of. It is difficult to overestimate the impact of these results on our views of nervous system behavior.

Meanwhile, an exactly analogous situation has been arising in cell biology with the growth of our knowledge of biochemical control mechanisms. Here it seemed more natural to begin with a continuous-time description where state variables are concentrations of specific chemical substances whose interactions according to the rules of chemical kinetics lead immediately to equations of the form (1). Here again (cf., e.g., Kacser 1957) we could build continuous networks which have properties reminiscent of real cellular systems. This work was greatly spurred by the discovery of a functional unit analogous to the neuron, called the *operon* by Jacob and Monod (1961), and the discovery that these units could interact in excitatory and inhibitory fashion (called induction and repression respectively) with threshold phenomena also appearing naturally (Rosen 1967). Sugita (1961) implicitly pointed out (what could be guessed from what has already been said) that independent discrete-time description of biochemical control could be given. This gave rise to the notion of a *biochemical automaton* (Apter 1966). An idea of the degree of the analogy between biochemical networks and the central nervous system can be gathered from the fact that the networks proposed by Jacob and Monod as models for differentiation are essentially identical with those proposed 30 years earlier by Rashevsky and Landahl for learning and discrimination. (It is important to note, however, that whereas the neural theory of the central nervous system is intended as an exact expression of the cell theory for that system, the corresponding operon theory falls below the level of the cell in a biological sense; moreover, whereas the model neuron is, as the name implies, intended to correspond to a structural unit, the operon refers to a functional unit rather than an anatomical one. However, as we shall see in terms of automata-theoretic descriptions, this distinction disappears on further analysis.)

Having now identified the units of organization of two different but analogous biological systems, and having specified in a general way not one but two kinds of system description appropriate at this level, it is time to inquire whether, and how, we may use these as a basis for a discussion of hierarchical organization in these systems. To see what is required, it is now time to inquire more deeply into what conditions must be satisfied by the system descriptions at this level. To do this we will now briefly describe the only example known to me in which the problem of hierarchies has been at all successfully solved. From this we hope to draw at least some of the necessary conditions we seek.

A SUCCESSFUL SOLUTION TO THE PROBLEM OF HIERARCHIES

The example I wish to consider is that of a thermodynamic system; more particularly the kind of system which we call a gas. We all know that we can regard a gas in two quite different ways. On the one hand, we can consider a gas as a system consisting of a large number of individual (Newtonian) particles, while on the other hand, a gas may be considered phenomenologically, simply as a continuous fluid without any substructure. Each of these viewpoints requires a different kind of system description. If we consider the gas as a system of particles, then we know from the theory of Newtonian dynamics that the state variables appropriate to describe the system are the coordinates and momenta of the individual particles of the system; this corresponds to what we usually call the *micro-description* of the system. If we view the gas simply as a fluid, then the appropriate state variables are quite different; they are the thermodynamic variables – pressure, volume, temperature, etc. These correspond to the *macro-description* of the system. Which description we use depends entirely on which aspects of the behavior of the system we are interested in; i.e., on the way in which *we are interacting with the system*. We have before us, then, a true example of a hierarchical system, one in which two apparently independent system descriptions are appropriate to different activities being simultaneously manifested by the same system.

We have said that the basic problem of hierarchical systems is to relate the appropriate description to the various hierarchical levels, and more particularly, to determine the extent to which the lowest-level description determines the properties of the system at the higher levels. In the present example there are only two levels, the lower described by the micro-description, the higher by the macro-description. The solution of the basic problem requires a tool which will enable us to relate these two descriptions, and as you are all aware, such a tool exists; it is called *statistical mechanics*.

Let us briefly illustrate what this tool looks like and how it works; this will be important to our later developments. We want to begin with the micro-description, and work upward to the macro-description. What is the micro-description? We know that the fundamental state variables are the displacements and momenta of the individual particles which make up our system. According to Newtonian dynamics, the kinetic properties of the system are given by the *equations of motion* of the system which express the momenta as functions of the state variables. This leads to a system of first-order simultaneous differential equations which, when integrated, determine all possible motions of the system; any individual motion of the system is specified when the initial values of the state variables at a particular instant are given.

The really important thing to notice about the equations of motion, and the machinery for setting them up, is that *they are supposed to contain all the information which can ever be obtained about the system*. This is one of the basic postulates of Newtonian dynamics; knowing the state variables at one instant and the equations of motion, we are supposed to be able to answer any meaningful question that can be asked about the system *at any level* (or at least, at any higher level). Of course, this answer may be *wrong* to the extent that Newtonian dynamics provides an incorrect description of the system at the microlevel; but this does not concern us here. What is important is that, whatever the underlying microdynamics, it must be postulated at the outset to have a universal character with respect to all higher levels.

This initial postulation of universality seems to solve the problem of hierarchies by tautology, but it really doesn't; for we must *supplement* this postulate by giving a specific recipe for answering higher-level questions in terms of the micro-description. And Newtonian mechanics does this. The relevant data here are contained in appropriate state functions, or *observables*, of the system. Answers to questions about individual states of the microsystem are obtained by evaluating the appropriate observables on those states; answers to questions about *families* of states are obtained by some kind of operation on the values assumed on the states of the family by the appropriate observables. And in classical mechanics, once the state is known any observable can be evaluated.

In particular, the universality postulate implies that *thermodynamic* questions can be answered in terms of the underlying micro-description. How? The state variables appropriate to the system at the *gas* level must be expressed in terms of the observables of the underlying mechanical system; i.e., as functions of the state variables of the microsystem. It is then a question of finding the appropriate observables and the operations which must be performed on them. This is what statistical mechanics does. It identifies a thermodynamic state (macrostate) with a *class* of underlying microstates, and then expresses the thermodynamic state variables as averages of appropriately chosen micro-observables over the corresponding class of microstates. The equations of motion of the underlying microsystem are then inherited by the thermodynamic state variables, allowing us to express in principle (through in practice only as yet in very special situations) the kinetic behavior of the system at the macrolevel.

REQUISITES FOR LEVEL DESCRIPTION

To sum up: to solve the problem of hierarchical organization in this simple case, we need:

1) The universality of the underlying microdynamics, which assured us that we could express *any* aspect of system behavior in terms of the micro-description in principle.

2) A determination of how the state variables of the macro-description could actually be described in terms of the

microdynamics; i.e., in terms of the observables of the microsystem. This determination is highly non-trivial, and without it, the universality assumption is operationally vacuous.

3) The implementation of 2) to actually derive the kinetic properties of the macrosystem from those of the microsystem. In this example, (2) and (3) are what statistical mechanics does.

There is yet another aspect to the problem of hierarchies in thermodynamic system which must be explicitly recognized. The unique role of statistical mechanics in physics (i.e., that of effectively relating two levels in a hierarchical system) has tempted a number of authors, working in a variety of biological areas, to try to mimic the statistical mechanical formalism in order to get some insight into particular hierarchical structures in biology, including the central nervous system. These attempts have not been very successful, and it is important to inquire why. The fault does not lie in the formalism itself, which does not depend on the underlying dynamics, and should in principle work regardless of whether the dynamical equations are those of Lotka-Volterra populations, biochemical control systems, or model neurons. The difficulty becomes clear when we consider the historical order of ideas in the development of statistical mechanics in physics. The development was as follows: *first* came the gas laws; i.e., the phenomenological specification of *macrosystem* behavior and the determination of the state variables appropriate to this specification. *Second* came the specification of the *microsystem* dynamics, and *last* came the statistical mechanical formalism connecting the two. This order is crucial; I feel secure in asserting that if the gas laws had not been known *first*, they would never have been discovered through statistical mechanics alone. Formalism will indeed enable you to form any averages you want, but it will not tell you what these averages mean, and which of them are useful and important in specifying and describing macrosystem behavior.

In all attempts to apply the statistical-mechanical formalism to biology, however, this historical order has been permuted:

first has come the formalism, imported from physics; *second*, the microsystem description; and *last*, the attempt to apply the formalism to the microsystem description to generate information about the macrosystem. It is obvious now why this information is not readily forthcoming. The fault lies in our initial ignorance of system descriptions at the upper levels of the hierarchy, and the basic fact that statistical mechanics *alone* cannot decide the relevance of any particular average to the macrosystem description.

Let us sum up the thrust of what we have done so far. According to the very definition I have proposed for hierarchical organization, the first essential point is that *no one type of system description can possibly display by itself a definite type of hierarchical structure* for the simple reason that we recognize such structure *only* by the necessity for different kinds of system description at the various levels in the hierarchy. But, as we have seen in the case of Newtonian mechanics, certain kinds of system description (universal ones) contain, in principle, information which can be identified with system descriptions at higher levels. This property is clearly the first requisite of a theory which purports to deal with a hierarchically organized system. There are, as we have seen, two other requisites independent both of the first requisite and of each other, namely: a supplementary procedure for isolating this information in a convenient form (a role played by statistical mecahnics in our example), and a decision procedure for identifying the relevant information with system properties at the upper levels (a role not filled either by the lowest-level description or by statistical mechanics, and on the basis of our experience so far seems to require a previously formulated independent description of the upper levels).

Next, we shall view automata theory as a candidate for a lowest-level system description. Our discussion thus enables us to narrow our goals. We concern ourselves here with the *first* requisite mentioned above; does automata theory, as a system description, contain information which pertains in principle to higher levels in a hierarchy? Is automata theory universal?

Let us now, having completed our extensive detour, return to the problem of hierarchies in cell biology and the central nervous system. As we pointed out earlier, each of these systems admits, at a particular level, two independent kinds of system description – a continuous-time and a discrete-time description. Our discussion of statistical mechanics only pertains to continuous-time descriptions (although the three conditions drawn from that discussion are generally valid, and pertain to all kinds of system description). We have seen that with regard to continuous-time descriptions of the neural and operon theories, two of the three conditions for hierarchical activity are satisfied; but in order for effective hierarchical structures to be generated, we must supplement these descriptions by external information about what consitutes a state description for the system at other (higher) levels in the hierarchy.

AUTOMATA THEORY DESCRIPTION: CANDIDATE FOR HIERARCHICAL STRUCTURE

What is the situation for automata-theoretic descriptions? It will be convenient to recast the discrete description form in terms of *sequential machines*. The theory of sequential machines is generally formulated in abstract axiomatic terms as a set of data and the relationships between them. These take the following familiar form; we are given:

1) A finite set of states Q, a finite set of input symbols X, a finite set of output symbols Y.

2) A next-state map $\delta: Q \times X \rightarrow Q$.

3) An output map $\lambda: Q \times X \rightarrow Y$.

These determine the sequential machine $\Delta = (Q,X,Y,\delta,\lambda)$. Note that the data defining the sequential machine are purely abstract entities. We can *realize* these data in a variety of ways. In particular, we can associate the abstract set of states Q with the states of a particular modular network, and thus obtain a *realization* of an abstract sequential machine in terms of a modular net. Indeed, given any abstract sequential machine, there is a network (built, say, out of McCulloch-Pitts neurons) which will realize it (Arbib 1964). There is no reason why such a realization should be unique, and indeed, in general each abstract sequential machine will determine a *class* of modular networks, each of which will be a realization of the abstract machine. These realizations will differ in their specific structure, but will all be functionally indistinguishable; to use another word, they will be *analogs* of one another in the sense in which we have already used this term. Thus we can study a sequential system either entirely in the abstract, or in terms of any one of the analogous networks which realize the abstract machine.

These considerations point up the fact that there is a fundamental ambiguity in automata theory which, depending on one's point of view, is either a basic strength or a basic weakness of the theory in describing the activity of the central nervous system. I believe it is fair to say that most workers in neural networks are seeking to establish *isomorphisms* between model networks and their biological counterparts; correspondences which will identify each formal neuron in the network with its counterpart biological neuron; each formal synapse with a biological synapse, and conversely. But "the real" network we seek is but one of a class of functionally identical networks; and if we work (as we always do) from functional properties of networks, how can we extract "the one" we seek? On the other hand, the theory also tells us that, if we want to deal with *functional* properties of a network, we can learn a great deal by working with any convenient analog which exhibits these properties, or even entirely in the abstract with a sequential machine. We shall return later to the question of whether it is of interest to seek such structural isomorphisms, and if so, how to go about it.

We now turn to the question of how the theory of automata is in fact used to describe the activities of, say, the central nervous system. Since we have seen that a particular modular net can always be replaced by a sequential machine, and if we assume (as we always do) that the central nervous system is a modular net, we may as well work with the more general and convenient sequential machine terminology.

It looks at first sight as if the sequential machine formalism is quite different in principle from other kinds of system description; for instance, from the kind of description envisaged in the Newtonian description of a mechanical system. But closer scrutiny reveals that this is not the case, and in fact, that the sequential machine formalism is really a close paraphrase of Newtonian dynamics in a discrete setting. The set of (discrete) states Q plays exactly the same role as does the phase space of Newtonian dynamics; the set of inputs X plays exactly the same role as do the forces applied to a mechanical system; and the next-state map δ plays exactly the same role as do the dynamical equations of motion. In each case, knowing the initial state and the forces acting on the system (i.e., the input sequence to the sequential machine) we can predict the entire temporal course of the system.

In other words, the description of a modular network as a sequential machine is really the same kind of animal as the description of a mechanical system in conventional dynamical terms; the differences between them are purely technical and do not involve matters of principle. This is a key observation, for it is now tempting to suppose (matters of technique aside) that we can do anything with a purely dynamical description. In particular, the question now is: can we use an automata-theoretic description of a modular net as a micro-description, from which we can effectively generate descriptions of hierarchical activities in the biological systems in which we are interested?

As we have seen, an automata-theoretic description of a modular network is in principle very similar to a conventional dynamical description. We have already seen how a dynamical description could be made to yield information concerning higher levels in a hierarchy. It thus remains to see whether the automata-theoretic description can likewise be made to yield such information. Indeed, if automata theory is to provide more than a very partial picture of biological activity, this must be the case.

We have seen that the first essential for a hierarchical theory is that the system description lying at the lowest level in the hierarchy be *universal* in the sense that it contain all the information which can be obtained about the system at any higher level. In order that this postulate be not vacuous, we also need explicit recipes for extracting this information and ascertaining its relevance; i.e., for expressing higher-level system descriptions in terms of what is happening at the lowest level.

What is the vehicle for this information in the dynamical description? The vehicle consists in the enormous set of *observables* of the system which is at our disposal; i.e., all real-valued functions on the state space of the system, together with all the operations that can be performed on them (e.g., averaging over classes of states). This is in fact the specific machinery in which resides the effective universality of Newtonian mechanics.

Let us look now at the parallel we have drawn between dynamical and automata-theoretic descriptions. These parallels involve the respective state spaces and the equations of motion. The observables of the dynamical description are numerical functions on the state space, all explicitly visible in the dynamical description. But when we turn to the automata-theoretic description, what do we find? We find but a single observable; namely the out-put function of the machine. To get another observable, we need another machine. In other words, the theory of automata has paraphrased everything about dynamical descriptions *except the explicit machinery required in order for the dynamical description to have a universal character.*

Exactly the same shortcoming arises if we retreat to the modular network level, except that in this case we have a few more potential observables as a consequence of the specific structure of the network. Here, the output function of a modular net is obtained by a specific choice of neurons called output neurons. If we choose these differently, we automatically get a new output and a new output function, which is, of course, an observable of the network. Thus, a particular modular network can give us a certain finite number of observables instead of the single one we obtain in the sequential machine formalism. But it is clear that universality requires the full richness of the set of observables of the system; no sample (certainly no finite sample) will do.

From these considerations, it appears very much as if there can be no question of hierarchical structures being based on automata-theoretic descriptions, even in principle, since the mechanism to generate higher-level descriptions is explicitly abstracted out of the description right at the outset. In other words: though we may increase *complexity* in a modular net by adding more and more units interconnected in richer and richer ways, this increase in complexity cannot by itself, elevate us into upper hierarchical levels. The machinery by which we might have hoped to do this is missing.

But if we have come this far, we have probably come far enough to observe that all is not lost. We may yet retrieve a possibility to account for hierarchical structures in automata-theoretic descriptions by attempting to re-introduce the required machinery. We know what is missing: more observables. An observable is a function on the state space; it is a *computation.* But automata and modular nets can do nothing if not compute; therefore we can retrieve any particular observable we may want by allowing an appropriate computing automaton access to the state set of our original system. In this fashion we may hope to regenerate, at least formally, the groundwork for a universal micro-description, the first prerequisite for hierarchical organization.

In fact, this is exactly what is done in network models which do exhibit some kind of hierarchical activities; i.e., explicit behavioral properties at a higher level. For instance, in the central nervous system we have the cell assemblies of Hebb (1949); the various pattern-recognizing devices, such as the perceptron of Rosenblatt (1962) which appear to make abstractions; and the heuristic problem solvers. But these are not *simply* modular networks; they are modular nets to which further structure has been appended. And this further structure serves precisely to generate new observables in terms of which the higher-level activities may be (at least partially) characterized. For instance: most of the above-mentioned models involve a totally new idea: *facilitation*, the manipulation of threshold or connection strength. Facilitation is an idea utterly foreign, by itself, to automata theory; it is a new construction principle, which can be *reduced* to conventional automata-theoretic terms only by the appending of new observables. A net with facilitation can be replaced by a (much bigger) net without facilitation; the difference in size between the two networks is due to the incorporation of new computing apparatus to generate new observables and use them as the basis for describing the higher level activities.

All this is in accord with what should happen in automata-theoretic descriptions in terms of the parallels we have drawn between these descriptions and conventional dynamics. We also see something else happening, again expected in terms of our parallel, but one which has a bearing on our *interpretation* of modular network theory in terms of the central nervous system. It is perhaps most easily stated initially in dynamical terms, and it is this: when we have succeeded in building the state variables of a lower-level description, we may use these new state variables, and the equations of motion they inherit, as the micro-description for the next level up in the hierarchy. The *forms* of these successive micro-descriptions will remain very much the same; only the *interpretations* vary. In the study of biochemical control systems, we start phenomenologically with a dynamical description in which the state variables are concentrations, and where the equations of motion contain certain "rate constants". These "rate constants" can themselves be derived from statistical mechanical arguments of the kind we have sketched or from a set of similar-looking equations, but where the state variables now describe the specific structure of the molecules involved. Likewise, in the automata-theoretic description, a net with facilitation behaves very much again like a modular net, *except* that we must now interpret the modules as being *assemblies* (in Hebb's terminology) of the original modules at the lowest level. In other words, we may use the automata-theoretic description independently for each of the several levels of the hierarchies of our system, just as we do in the dynamical case. The interpretations (i.e., the state descriptions) change, but the descriptive machinery is preserved. This is simply a restatement of the remark of Bradley mentioned earlier.

All of this reduces simply to the conclusion that, if we want to build a hierarchical organization on the basis of an automata-theoretic description at the level of a functional unit (i.e., the neural or the operon), a desire implicit in all of neural network theory, then in particular *we must abandon our attempt to find a structural isomorphism between neural network models and the central nervous system itself*. We must do this because, while the real neurons have all of their observables built into them as dynamical systems, and hence do in principle allow the generation of hierarchies, the automata-theoretic description only allows us one (or at best a few) observables at a time. This, I think, is the deeper significance of McCulloch's assertion that his formal neurons are *impoverished.* It is true, as we have seen, that we can build back into the network any particular observable we want, using any particular kind of module (e.g., a McCulloch-Pitts neuron or other kind of neuromine) *but this requires extra modules* which have no *structural* counterpart in the real network. There is, of course, a *functional* counterpart to the supernumerary modules, but this functional counterpart is buried inextricably in the incredibly rich dynamics of the real neuron. It is precisely here that the possibilities for the generation and representation of hierarchical organizations resides, and it is precisely this aspect which has been, as a matter of principle, systematically neglected in automata-theoretic descriptions of centeral nervous system activity.

The above analysis, as I said before, is only a beginning investigation and deals only with the very first requisite for building a theory of hierarchies in automata-theoretic terms. It does not tell us what kind of observables we should build into modular networks to allow for particular kinds of hierarchical behavior. It does not tell us what operations to perform on these observables in order to put them into a useful form for higher-level studies (that is, it does not provide an automata-theoretic analog to statistical mechanics). It does not tell us which of this higher-level information is *relevant* as a system description at the higher levels. All this it does not do. And yet, I believe that this analysis may serve to recast the problem in a more manageable form – one in which conceptual difficulties are replaced by technical ones. And as one deeply interested in hierarchical organizations in biological systems, I hope that this analysis, or some like it, will represent a forward rather than a backward step in our struggle to understand these systems.

CONCLUSION

I would like to stress once more the analogies between cellular biology and the biology of the central nervous system, which are the basis of many other similar analogies in biology for which the principles developed above hold. Such analogies must and should be systematically exploited, just as they are in analog computation; for the dynamical characteristics of any pair of analogs, in this sense, are the same. The exploitation of such analogies between the central nervous system and cellular control systems seems valuable to me for a variety of reasons; among them:

1) The conceptually unifying effect that will follow from the discovery of principles governing the behavior of such important but structurally dissimilar systems. This is a unification which seems, to me at least, far more fruitful and satisfying than any which can be derived from a simple reductionism. Such a unification is already visible in the organization of physics and around variational principles, as opposed to the reductionist attempt to derive all of physics from the theory of elementary particles or unified fields.

2) The interplay of intuitions developed on two analogous but physically highly dissimilar systems cannot help but be valuable to both. For the moment, most of the flow seems to be from the direction of the highly developed theory of modular nets in the central nervous system toward the theory of biochemical control (especially in connection with differentiation, because of the *constructive* aspect of the theory, the actual building of networks with specific logical capacities). But the flow is bound to go the other way as well, particularly in connection with hierarchical activities which are perhaps more clearly visible and quantitatable in cellular systems than in the central nervous system.

3) The technological applications of experience we can gain in the study of alternate realizations of a common functional organization. The analogy we have indicated points up the fact that we can construct modular networks with hierarchical properties, out of biochemical components in at least two essentially different ways – one in which the modules are operons, and one in which the modules are neurons. There may be other ways as well, perhaps involving artificial engineering components; the synthesis of such systems is one of the goals of both the area of bioengineering and of biologists themselves.

There may likewise be realizations built on other chemical systems than one familiar $H_2O - CO_2 - N_2$ system; the investigation of this possibility should have a concrete bearing on exobiology. There must indeed be realizations corresponding to eobiotic forms long extinct, the investigations of which will make possible a direct attack on problems pertaining to the origin of life.

It seems to me that it is most fruitful to approach this as a problem of alternate realizations of functional properties than through the study of purely physical characteristics which, as we have seen, need not be simply related to functional behavior, particularly at the higher levels.

REFERENCES

Apter, M. 1966. *Cybernetics and Development.* New York: MacMillan.

Arbib, Michael A. 1964. *Brains, Machines and Mathematics.* New York: McGraw-Hill.

Bradley, D. F. 1968. "Multilevel Systems and Biology." In *Systems Theory and Biology*, ed. M. D. Mesarovic, pp. 35-58. New York: Springer-Verlag.

Elsasser, W. M. 1966. *Atom and Organism.* Princeton, N. J.: Princeton Univeristy Press.

Hebb, D. O. 1949. *The Organization of Behavior.* New York: John Wiley & Sons.

Jacob, F., and Monod, J. 1961. "Teleonomic Mechanisms in Cellular Metabolism, Growth and Differentiation." In *Cold Springs Harbor Symposia on Quantitative Biology*, Vol. XXVI.

Kacser, H. 1957. Appendix to *Strategy of the Genes* by C. Waddington, pp. 191-249. London: George Allen & Unwin.

McCulloch, W., and Pitts, W. 1943. "On the Ideas Immanent in Neural Activity." *Bull. Math. Biophys* 5:115-33.

Polayni, M. 1968. "Life's Irreducible Structure." *Science* 160:1308-12.

Rashevsky, N. 1938. *Mathematical Biophysics.* Chicago: University of Chicago Press.

Rosen, R. 1967. "Two-Factor Theories, Neural Nets and Biochemical Automata." *J. Theoret. Biol.* 15:282-97.

Rosen, R. 1968. "Some Comments on the Physico-Chemcial Description of Biological Activity. *J. Theoret. Biol.* 18:380-86.

Rosenblatt, F. 1962. *Principles of Neurodynamics: Perceptrons and Brain Mechanisms*. Washington, D. C.: Spartan Books.

Sugita, M. 1961. "Functional Analysis of Chemical Systems *in vivo* Using a Logical Circuit Equivalent." *J. Theoret. Biol.* 1:415-430.

■

Theorems on Boundaries in Hierarchical Systems

John Platt*

Simon (1962), in his beautiful paper on "The Architecture of Complexity," has discussed the structure and assembly of 'nearly-decomposable systems.' These are systems which can be broken up (in thought or analysis) into sub-systems such that the interactions *within* the sub-system are relatively strong and numerous compared to the interactions *between* the sub-systems. Deutsch (1953, 1956, 1966) has shown quantitatively how the high level of interaction-rates within a nation, as measured by the flow of mail or communications between cities, changes to a lower interaction-rate between nations. Miller (1965) has discussed the types of different sub-systems which are needed to make a viable organism, or system, at a given level. His sub-systems include boundaries (functional rather than mere mathematical surfaces), and input and output transducers to handle flows of materials, energy, and information. These are in addition to the more commonly emphasized sub-systems such as processors of materials, energy, or information (stomach, brain), storage units (fat deposits, memory), reproductive sub-systems, and so on. Plessner also emphasized the functional character of boundaries for transactions with the environment in his concept of "positionality," as discussed recently by Grene (1968, pp. 76-78).

But in all this, a detailed analysis of the unique role of the properties of boundaries between a system and its super-system, or an organism and its environment, has tended to be neglected. It is interesting to think about some general properties that boundaries must necessarily have, or general theorems that they must satisfy, whatever their level. These theorems can generate some additional "testable cross-level hypotheses" to be added to the lists of such hypotheses that Miller has given for living systems.

**Mental Health Research Institute, University of Michigan, Ann Arbor, Michigan 48104.*

In the theorems suggested here, I especially have in mind applications to a very general model of a biological or living organism or a sophisticated automaton, a model that could be described as a *sensory-motor parallel-processing decision-system*. For brevity it could be called a "Seymour" system, if one allows the pun on "see-more." Since it is defined as parallel-processing, such a system has to be a multiple- receptor multiple-effector amplification system which uses a total pattern of sensory inputs to decide among a set of alternative motor outputs. (The outputs must be alternative, because a system would not be viable if it had multiple dependent motor outputs which would be in conflict much of the time.)

For a Seymour system to represent a living organism or a fairly independent automaton (such as the Surveyor series of devices on the moon), it should have the following additional characteristics.

1) It should operate (i.e., make decisions) in real time.

2) It must be an individual (undivided) to be effective (i.e., it must make non-conflicting motor decisions).

3) It probably must be bounded (i.e., with a clearly definable internal-external separation-surface which is closed).

4) It must lose information in making decisions–since an operating decision involves selection and amplification (Platt 1956). This means it must group the possible stimulus-fields into output-equivalent classes or patterns–which corresponds to the abstraction of properties.

5) It should be placeable on a scale somewhere between a 'non-learning' and a 'learning' system, according to that flexibility in modes of response, or *a priori* classifications, built into the original structure of the system.

With this kind of sensory-motor decision-system in mind as being of special interest to us, we can go on to see what theorems can be found about boundaries in general and in this particular case.

THEOREMS ON UNDIFFERENTIATED BOUNDARIES

I. Boundary Definition. A sub-system in a nearly-decomposable system in n dimensions will have boundary surfaces of $n-1$ dimensions between a high-interaction region and a low-interaction region. The surface may be taken as passing through a family of points where some parameter such as "interaction-density" has a maximum gradient. The continuity and smoothness of such surfaces will be an interesting question, but it seems likely that in real physical systems, in the continuous approximation, the surface will be smooth and continuous except for a few singular points. In a telephone system, of course, the boundary of the system could be taken as the set of all telephones.

II. Coincidence of Boundary-Surfaces of Different Properties. This theorem seems to have been first pointed out by Campbell (1958, 1965). The boundary-surface for one property (such as heat-flow) will tend to coincide with the boundary surfaces for many other properties (such as blood flow, sensory endings, physical density, and so on) because the surfaces are *mutually-reinforcing.* I think this somewhat astonishing regularity of nature has not been sufficiently emphasized in perception-philosophy. It is this that makes it useful and possible for us to identify certain sharply-defined regions of space as "objects." This is what makes a collection of properties a "thing" rather than a smear of overlapping images.

The requirement of boundary-coincidence that is built into our neural interpretations of the world is obviously useful for survival in a world of this type, and it gives "true" observations a multiply-determined character, with confirmations, which lies, illusions and dreams do not have. Conversely, any violations of boundary-coincidence have an upsetting fascination for us, as in tales of ghosts, which can be seen but not touched, or in the case of glass and mirrors which can be touched but not seen. Even a dog will bark at the biological contradictions of the mirror. The requirement of boundary-coincidence may be a major mechanism in Held's reafferent-stimulation adaptation to visually distorting glasses (Held 1965), where the failure of self-motions of the hand to be at the point or object indicated by the (distorted) visual system gets compensated for very quickly. This may also be the neural requirement that gives us a "body-image" of ourselves, continuing to provide us with a "phantom limb" after an amputation (Simmel 1958, 1962), a limb which however "shrinks" over the weeks until it fits the new body contours as seen from the outside.

III. Parallel and Perpendicular Law. All gradients and flows in the region very near to a boundary will tend to be either parallel or perpendicular to the boundary. This is because flows of any kind will be down minimum or maximum lines (ridges or valleys) of some property; and the perpendicular line or "normal" to a surface, or the parallels to the surface (normals to the normal), form the only unique set of minima or maxima of any property-gradients. In the case of a biological organism, the flows along the normal to the surface would include heat flow, perspiration (average, through the skin), osmosis, and sensory inputs. Flows tangential to the surface would include blood flow in capillaries, surface stretching and contraction (surface tension), muscular contraction, and so on. Evidently in the region close to a boundary, the strong-coupling interactions *within* the system are parallel to the boundary, while the weaker-coupling interactions *between* the system and the larger super-system flow in and out perpendicular to the boundary.

THEOREMS ON DIFFERENTIATION OF BOUNDARIES

IV. Gates. The $n-1$-dimensional boundary surface may likewise have lower-dimensional boundaries in or on it. A mouth is the most obvious example. The $n-1$-type boundaries will be bounded by $n-2$-dimensional lines or lips. These might be peripheral, as in the case of the edges of a jelly-fish or a sand dollar. But in looking at the system more generally as being an identifiable compact region of space, the $n-1$ boundary will be nearly closed, and the lips will form closed loops around "gates" or "mouths," which will be centered at unique $n-3$-dimensional points on the surface. Flows through gates may be of materials, energy, or information. The gates will commonly be specialized for certain flows to or from the larger external environment. In the telephone system, perhaps we should think of the boundary as being made up entirely of gates, with each telephone being a gate.

In comparing systems of different sizes, we see that certain flows or transactions with the environment which are carried on over the whole boundary surface at one level of sizes, come to be carried on through specialized gates at another level of sizes. Certain flow transactions are organizationally 'cheaper' when spread over the whole surface uniformly; for example, thermal radiation. Others are cheaper when localized; for example, thermal convection, which, when the temperature gradient passes a certain critical level, gets concentrated into localized air columns from a biological body as well as from the earth (Benard cells, thunderstorm updrafts).

The question as to whether a distributed or locally concentrated mode of flow occurs depends on the absolute size of the system, and also on rates of demand, gradients, feedback controls, and so on. Thus thermal transport through a liquid, or oxygen transport through the epidermis of a small creature, may be uniform over the surface when the system is small and thin, but may tend to localization (as through a trachea) when the system gets larger. The entry of drugs or arms into a nation may be across a whole boundary, by smuggling, until controls are imposed which force them to come through a few Customs Ports. Obvious examples of gates include: at the microscope level, sensory receptor cells, sweat glands, and so on; and at the macroscopic level, the mouth, eye, ear, anus, and so on. (The 9 openings and the 13 secretions, as the Chinese say.)

Lemma 1. Many gates may coincide. This occurs because of mutual reinforcement, or economy, as in Theorem II. A good example is the mouth, which is a gate for food, water, air, and speech.

Lemma 2. For some processes, multiple similar but spatially discrete gates will be needed. This makes a 'mosaic receptor,' whose pattern and spatial separation, *d*, or degree of spatial resolution, are limiting filters on what the system can take in and "know" about the external world. Examples are the retina, the cochlear spiral – where the spatial mosaic must determine in principle a frequence-mosaic – the array of touch receptors, and so on.

Lemma 3. Inputs to a given gate are in principle 'private.' If they are near the threshold level, there is no reason why they are necessarily observable through other gates in the same system, or through gates in other similar systems nearby.

If we define "public discussion" as meaning discussion between two biological systems using a public language and ostensive definitions in describing environmental features to each other, this last Lemma suggests that this public discussion may be limited to *field properties*, that is, to properties which do not change appreciably over distances large compared to the mosaic spacing, *d*, of input gates. Fields of this type might include light distribution (classical), sound, and chemical fields. Of course our neural networks interpret the regularities of these fields, structuring them as coming from a relatively few "sources" or "objects" in the environment. The "observable" object-dimensions in this restructuring are probably limited then to a range of scale neither too small nor too large compared with *d*. Our unaided visual resolution is limited to the angular spacing between adjacent rods or cones in the retina, and makes it impossible for us to see atoms; and conversely, large surrounding fields or constant background illumination tend to be cancelled out in visual perception.

Lemma 4. There are "gates" in time as well as in space. Goodwin (1963) has emphasized that any physical-chemical system, like a biological system, involving coupled differential equations, will in general not have steady-state solutions, but oscillating or periodic solutions. Periodic feeding, or excretion, or reproduction, may be energetically more favorable than continuous fluxes in many cases. Naturally these periodicities in time for different functions or flows will be mutually reinforcing and will tend to become locked in to each other, just as in the case of the mutual reinforcement of space boundaries. Note that these periodicities will also occur at the DNA level, as for read-out of an operon in making a particular protein at a given point in the cell's life cycle, as well as in the copying of the DNA preparatory to cell division.

Periodicity in time means pulsed inputs and outputs. When this is combined with the spatial periodicity of the gates, it means that each gate, to the extent that it is independent of its neighbors, will have an "on" time and an "off" time or refractory time. In the limiting case, this means 'unitary' – all or none – inputs or outputs. We see this not only in the all-or-none responses of the visual cells at the one-quantum level, but in the coupled periodicities of the heart muscle cells, with their all-or-none periodic contractions, and in the coupled daily peridocities of eating and defecating.

V. Transducers at Gates. The functional specificity of gates is not only to pass materials, energy, and information between the organism and the environment, but also to transform these entities back and forth between the external and the internal 'language' of the organism. At the boundary, thermal radiation from the environment is transformed into warmed blood. In the mouth, the salivary enzymes begin to transform the food into the substances needed for internal synthesis. But this theorem is particularly applicable to the information gates, where the transduction in the first receptor cells is localized and specialized. In general the internal language of a system is never the same as its exchange language to the environment or to other systems. Light, sound, and chemical information into our receptor cells must be transduced by specific molecules or structures into the electrical potentials or spikes of the neural network. Conversely for output, where the lung and throat muscles are transducers to the speaking voice.

ORDER-OF-MAGNITUDE RELATIONS FOR GATES AND NETWORKS

VI. "Surface-Volume" Relations of Networks. The "surface," or number of input gates, into a volume-filling network system, in general has a lower dimensionality than the number of nodes in the network itself. This is the analog of Theorem I for the case of a network with discrete nodes instead of a continuous volume. What this "lower-dimensionality" means will depend on the principles of intereconnection in the network. In the case of the nearest-neighbor connection of random points, like the lines connecting the corners in a mass of soap bubbles, a closed surface of approximate diameter D will cut of the order of $(D/d)^2$ lines ("gates" through the surface), while it encloses a number of the order of $(D/d)^3$ nodal intersections, where d is the mean length of the nearest-neighbor lines. For a large parallel-processing nervous system like a Seymour system, the number of input gates may be orders of magnitude lower than the number of nodes or synaptic connections in the network into which they feed, or the number required to make the necessary or possible discriminations among their patterns.

To put it another way, the number of interconnections in the network needed to make sense of the external world is probably necessarily greater than the number of sensory cells that are interconnected. If we have 4 photocells – or 4 eye-spots in a primitive animal – there are 16 on-off combinations between them, and it probably takes many more than 4 interconnections to distinguish these. If each node combined inputs from 2 nodes of the preceding stage (in a precessing system with successive stages), it would take $n = \log_2 m$ stages between m sensory inputs and any final simple decision among m alternatives, giving a network with the order of *nm* nodes, which would be much larger than *m* (Platt 1956). Correspondingly, in a human being, there seem to be no more than of the order of 10^9 sensory cells in the eye and ear and other receptors, but there are some 10^{11} neurons and perhaps of the order of 10^{14} synapses interconnecting them in the brain.

One nematode now being studied by workers on the nervous system is said to have just 59 neurons. It would be interesting to examine the relation between numbers of sensory inputs and numbers of neural interconnections from such an animal upwards to man. It seems possible that there might be similar theorems for the chemical input signals passing into a cell, which may pass through particular discrete "gates," or permeate molecules, in the cell wall membrane. In this case, however, it is not clear how to specify the chemical nodes or processing points or decision-points within the cell, but the problem is worth further examination.

The theorem also may have some value for larger systems such as a military organization with inputs from patrols or intelligence; or an economic network, or a national political system with foreign or border inputs; the difficulty with applications of this type would be in getting an unambiguous definition of the internal decision-points or nodes beforehand.

But if this theorem is correct for a biological neural network or a Seymore system, it leads to an important biological "indeterminacy," which I call "complexity-indeterminacy" (Platt 1966a, 1966b). For it would mean that I do not have early enough sensory cells to determine the state of every neuron in your brain at any given instant, even if I had enough microscopes and could use all my sensory cells for this purpose or could distinguish their 10^9 separate reports (as by giving an output-report on each of them). So even if I should believe and prove that any groups of neurons I study in your nervous system are completely deterministic in their behavior, I am faced with the paradox that I will not be able to make enough observations fast enough to prove that your whole brain acts in this way. Thus I can never find out whether your responses at any later instant are completely determined from your initial conditions and your sensory inputs or not. The totality of your interconnections is unobservable and the totality of your behavior is unpredictable in real time by any other human being, except statistically. Or as Ehrensvard (1965) has said, "Consciousness will always be one degree above comprehensibility."

This example makes us realize that our ideas of determinism and epistemology have been developed from the study of simpler problems. The physicist or chemist, with his 10^{11} brain and his 10^{9} sensory elements, observes and relates a very much smaller number of external variables, perhaps only 100 or 1000 or so planetary coordinates or chemical properties. The ideas developed in such cases may cease to be applicable, and certainly have to be completely reexamined, when it comes to the problem of trying to observe, predict, or know something of the same order of complexity as the observer himself.

VII. Relation of Number of Motor Outputs to Number of Sensory Inputs. For a viable living system, there is also probably a necessary relation between the number of sensory inputs and the number of motor outputs, so that the latter is comparable to the former, but may be smaller by something like an order of magnitude. One sensory input, on or off, can provide only enough information for one motor output, on or off; and m on-off sensory inputs likewise will provide just enough information to control m independent on-off motor outputs. On the other hand, sensory redundancy is valuable in overcoming degradation of the information, and it is fairly cheap in energy requirements compared to motor outputs, so that it may be an advantage for the sensory input patterns to provide appreciably more discrimination and internal checking than the motor outputs absolutely need. But on the other hand, even for m motor outputs, there is a reduction of information because of the requirement that they be non-conflicting. In a given interval of neuron pulse time, m *alternative* motor outputs would require only $\log_2 m$ bits of information, a much smaller quantity.

In support of this theorem is the observation that the number of motor neurons from the human brain is about one order of magnitude smaller than the number of sensory inputs, from comparison of the input and output fiber bundles. It may also be relevant that some experiments in which microelectrodes have been placed in the hand to pick up the electrical activity of individual motofibrils have shown that several human subjects were able to acquire voluntary control over every motofibril tested (Basmajian 1963). This requires that the network be capable in principle of fitting every motoneuronal output into a behavioral combination as needed. This would seem to be excessive specificity from an evolutionary point of view, if there were not enough sensory information available to make the choices among these behavioral combinations.

This is a complex subject, however, because one needs to know how to take into account the added input information which is acquired during cumulative learning. It will also require careful analysis to see how the problem is affected by the complexity of the organism (however that is defined) and its survival requirements. In a larger hierarchical system like an army, one sees dimly the possibility of similar relationships. Certainly there are many points where local input leads to local output, as when the platoon shoots at the snipers it has just flushed out (similar to a reflex arc in the nervous system); and there are probably numerical relations between efficient input-rate and output-rate numbers at each level of the hierarchy; with these relations being limited, of course by the input-output capacities of the individual human beings involved.

Some other conjectural theorems on time-relations between internode time constants, number of stages, and total response time in a parallel-processing decision-system are suggested in Platt (1956, pp. 187-9).

The theorems here, stated in this general way, are quasi-tautological because they are to some degree implied in the assumption of a hierarchical architecture of sub-system, system, and super-system, with transactional exchanges between levels. But they make some of these relations more explicit, and they are more than tautologies in that they suggest questions to study experimentally at the cellular level, the social group level, or the nation level. Some of the dichotomous questions – such as parallel versus perpendicular flows, or continuous boundary surfaces versus discrete gates – as well as the numerical values of gate spacings (d) and flow rates, might be accessible to model calculations. It will be interesting to see what other conjectural theorems and lemmas about boundary properties and behavior can be added to this list.

ACKNOWLEDGEMENTS

I am indebted to Drs. James G. Miller, John Burns, and Robert House for helpful discussion on a number of these points. This research was supported in part by PHS Grant GM 14035.

REFERENCES

Basmajian, J. V. 1963. "Conscious Control of Single Nerve Cells." *Science* 141:440-41.

Campbell, D. T. 1958. "Common Fate, Similarity, and Other Indices of the Status of Aggregates of Persons as Social Entities." *Behavioral Sci.* 3:14-25.

——. 1965. "Theories of Boundaries, Groupings and Systems." Ch 7 in *Propositions About Ethnocentrism from Social Science Theories,* eds. D. T. Campbell and R. A. Levine. Unpublished monograph, Northwestern University.

Deutsch, K. W. 1953. *Nationalism and Social Communication.* New York: John Wiley & Sons.

——. 1956. "Shifts in the Balance of International Communication Flows." *Public Opinion Quart.* 20:143-160.

——. 1966. "Boundaries According to Communications Theory." In *Toward a Unified Theory of Human Behavior,* ed. R. Grinken, 2nd ed., pp. 278-297. New York: Basic Books.

Ehrensvärd, G. 1965. *Man on Another World.* Chicago: University of Chicago Press.

Goodwin, B. C. 1963. *Temporal Organization in Cells.* New York: Academic Press.

Grene, M. *Approaches to a Philosophical Biology.* New York: Basic Books.

Held, R. 1965. "Plasticity in Sensory-Motor Systems." *Scientific Amer.* 213: 84-94.

Miller, J. G. 1965. "Living Systems." *Behavioral Sci.* 10:193-237; 337-411.

Platt, J. R. 1956. "Amplification Aspects of Biological Response and Mental Activity." *Amer. Scientist* 44:180-97.

——. 1966a. "Man and the Indeterminacies." *Perspectives in Biol. and Med.* 10:67-80.

——. 1966b. *The Step to Man,* pp. 147-49. New York: John Wiley & Sons.

Simmel, M. 1958. "The Conditions of Occurrence of Phantom Limbs." *Proc. Amer. Phil. Soc.* 102:492-500.

——. 1962. "Phantom Experiences Following Amputation in Childhood." *J. Neurol. Neurosurg. Psychiat.* 25:69-78.

Simon, H. A. 1962. "The Architecture of Complexity," *Proc. Amer. Philos. Soc.* 106:467-82.

■

Hierarchy, Entitation, and Levels

Ralph W. Gerard*

The overall plan of the symposium is obvious enough: the first day sessions looked at matter as such – at material hierarchies or systems and at formal systems; the second day, at living systems and at cosmic systems. My only regret is that somewhere along the line, social systems, which were in the original thinking, were more or less elided. But you can't have everything. The fantastic elevation and descent from 10^{42} to 10^{-42} magnitudes in the universe and the uncertainty at the two ends, makes me think of that lovely story of the evangelist who told his audience the world rests on a great rock. A hand went up, "What holds up the rock?" "Another rock," he replied and glared down his questioner. He completed his talk, but obviously still troubled, looked at his questioner and said, "And I want you to know it's rocks all the way down."

We have had many definitions of hierarchy here, most of them at least congruent, but not all identical. Let me try my own formulation in putting this together. I would say that the essence of hierarchy is in *subsumption;* the key is *sub* (or *super*). One can sub*ordinate,* which is a subsuming in the authority or control relation; one can sub*divide,* which is subsuming in the part and whole relation; one can deal with sub*species* or sub*routines* or sub*headings* in the relation of elements in a larger system; one can use the term sub*jacent* which is a kind of spatial subsuming; and so on.

In all of these there is an implication, of course, of the whole and the parts. The part may be related to the whole in terms of substance, in terms of time, in terms of space, in terms of influence, in terms of complexity, in terms of origin. One could go on and on and put other kinds of relationships in this part-whole subsuming. This has been presented from various viewpoints during this symposium and I shall not greatly elaborate. Here I would like to introduce certain

**Dean of the Graduate Division, University of California, Irvine, California 92664.*

neurophysiological examples to illustrate the notions of entitation – the identification of entity – and levels. I have presented some of my views in more detail elsewhere. (Gerard 1940, 1958, 1961a, 1961b, 1965.)

UNITS

The implication of whole and parts is, of course, that there does exist some kind of subordinate unit or units, some kind of a part; and a unit implies some kind of boundary. I think this is the essence of what Cyril Smith develops in his paper; at least, when we discussed it in the question and answer period, he allowed that discontinuity was perhaps more inclusive and to the point of his theme than is imperfection or irregularity; and a discontinuity is a boundary. Biologically, biopsychologically, boundaries reduce to edges. Surfaces have the utmost importance; without them, as Dr. Harrison points out, we couldn't even begin to structure the universe or any part of it. Making categories is man's great intellectual strength and weakness: strength, since only by dividing the world into categories can he reason with it; weakness, since he then takes the categories seriously.

The first effort to understand the universe is in terms of physical entities which are perceptually given to us. I will come back to that in another connection shortly, but now let me introduce you to a little more physiology, particularly neurophysiology; although Dr. Whyte referred to some of this work elliptically in mentioning Hubel and Wiesel.

The eye, which is our main receptor for informing us about spatial things in the world, is built with some very remarkable mechanisms for exaggerating boundaries or edges. One of the long-recognized points is that a visual boundary represents differences in intensity of light; whether the joining surfaces differ in color or brightness is immaterial, for intensity always enters. First, you are probably all aware of the fact that if one stares fixedly (and one has to fix the eye so that the usual jiggling Brownian-like movements of the eyeball are effectively cancelled) patterns vanish from consciousness within a few seconds. The slight movement of the retinal fixation point across the intensity boundary is essential to making it mean something. Next, in the work of Adian we find that if one makes a step increase in intensity of light, or almost any other stimulus, the frequency of messages coming up the optic (or other sensory) nerve, a series of blips of single nerve impulses will go way up but soon fall toward the initial level even though the higher intensity is maintained. So the movement of retinal receptors across the boundary between intense and non-intense regions is critical in giving a series of bursts of increased impulses (or bursts of decreased impulses, because this works in reverse in the same way). That is one mechanism for bringing out edges or structure.

Another was worked out more recently by Hartline. (Incidentally, both these gentlemen were given Nobel prizes for these bits of work.) Hartline studied the receptors of a relatively simple eye, that of the king crab. Each club-like ommatidium is clearly a separate photo-receptor unit, with a separate nerve fiber going from each to the brain of the animal. It had long been known that collaterals from these nerve fibers went laterally to connect with others, over quite a range. Hartlime showed that they are inhibitory fibers; that when an ommatidium is active, impulses in the collaterals tend to cut down discharges in the nerve fibers of neighboring ones. The effect tails off with distance between units. Incidentally, note here some kind of global property, not inherent in the properties of the individual unit. If one places an edge over the eye each unit on the bright side is firing at a high rate, and each unit on the dim side is firing at a low rate. But each bright ommatidium is strongly inhibiting its neighbors on both sides, and all are therefore firing less vigorously than if alone. But as the edge is approached, inhibition from neighbors is strong only from one side; on the dim side of the edge there is little activity and inhibition. So that the responses down each nerve from the bright region are greater as they approach the edge, where they are maximum. On the dim side responses are low and not much inhibited from fellow ommatidia, but those near the edge are more inhibited from the bright side. So the higher frequency of the bright area is increased near the edge and the lower frequency of the dim side is decreased; thus the frequency change across the edge is greatly exaggerated.

Well, these are some rather simple neurophysiological mechanisms that indicate how the organism distorts but does not falsify the world; perception does not give a homomorphic representation of the universe, but a distorted isomorphic representation. So much for edges and boundaries.

ENTITATION

Let us come back to the development of awareness of units. We start in terms of the development of the individual child, as he begins to structure his universe, and in terms of the race, as science has progressively structured its universe. First material units, more or less of human spatial and temporal dimension, are noted – objects that become nouns. Then one sees that these units may be made of subunits, and one dissects; this is morphology. Then one sees that what were originally recognized as separate units may fall into groupings, which is classification or taxonomy. Attention moves from the initial unit up to superordinate units and down to subordinate ones. Then with considerably more sophistication, one recognizes functional, and eventually developmental units as well as morphological ones – the whole history of science elegantly demonstrates this biology. I think that the important jump that must be made in most areas is from the morphological to the functional element, a point that Dr. Rosen at least touched upon.

Now, a word more about the initial recognition of elements. I have introduced – and I like the word – "entitation"; the identification of entity. I have sometimes annoyed my scientific collegues by asserting that entitation is vastly more important than quantitation. In science we tend to think that our great strength is our ability to quantitate, and of course it is what sets science aside from so many areas of human experience and activity; but it is perfectly meaningless to measure something, with higher and higher degrees of precision, if the thing you measure is more or less meaningless.

Perhaps one of the earliest historical examples of entitation was the primitive astronomy of combining stars into constellations. "Constellation," by the way, means putting stars together. These were phenomenologically of some interest; they were realistically – may I say genotypically rather than phenotypically – meaningless. So entitation is of the utmost importance. Many of the things that were presented to us in these two days had to do with the kind of entities one is trying to use in structuring a problem. I have always liked Whitehead's pithy statement, "Nature doesn't come as clean as you can think it." William James said it the other way, "The child faces the universe as a blooming, buzzing confusion." You have to structure it to be able to do something about it; but, as I also said earlier, you mustn't take the structure too seriously.

It's a bit like the umpires discussing their efforts. The first one said with some satisfaction, "Balls and strikes, I call them as I see them." The second, a little more arrogant, said, "Balls and strikes, I call them as they *are*." The third one, of greater experience and wisdom, said, "Balls and strikes, they ain't nothing *until* I call them." We are always calling them during life.

I get a little nauseated with the vulgarization of the term "breakthrough." At the moment, a local radio station is asserting with great self-righteousness that there has been a breakthrough in traffic reporting because it is using a computer to calculate travel time. A real breakthrough, scientifically at least, to me is when somebody has sufficient creative imagination – and courage to follow up, which may be even more important – to say "Let us look at the universe in terms of some new kinds of entities, some new kinds of units; or, what really comes to the same thing, in some new way of combining units"; because combining units gives a new unit at the superordinate level.

Again, many of you will remember Eddington's lovely statement, "We used to think if we knew one, we knew two, because one and one are two. We are finding that we must learn a great deal more about 'and'." This is the combinatorial problem and the essence of hierarchies; they are more than mere assemblies of units.

Interactions. This brings me, then, to the question of interactions of systems, of units in systems, in hierarchies. They always interact with each other at one level, what Dr. Wilson calls horizontal interaction; but they also interact up and down levels, including the whole. These vertical interactions are not symmetrical, but they also are not unidirectional. The whole influences the parts and the parts influence the whole. This has been a difficult notion to really come to grips with, so I was particularly pleased with Dr. Pattee's approach. At a very simple molecular level, where there is no mystery about it, he showed clearly such vertical influences from folding and even quantified them. It is in this interaction between the lower level and the upper level, the vertical interaction, primarily but maybe not exclusively, that I think the element of emergence comes in.

New global properties of new state variables have been reported by several; the folded molecule was one. Let me give some examples at the more complex biological level; you may not know the vast literature in experimental embryology. The fertilized egg divides into two, then four and eight cells, and so on, and the cells form certain layers which fold in or out until finally a newborn infant, of whatever species is formed. The embryo grows amazingly precisely, in spite of the errors that one sees in the newborn. The information necessary to specify all these morphological details, particularly the morphological connections within the brain itself, just could not be specified in the genes that are available. All the alleles of the 10,000 or 40,000 genes of man, whatever number you take, could not contain enough information; they could only specify certain programs that will produce particular kinds of changes.

Further, it is clearly not all program because one can interfere at any stage in the developmental process and, by simply making spatial readjustments, can change the destiny of a particular group of cells. For example, the early nervous system runs the length of the embryo as a folded tube and the early skin lies outside it as a kind of sac. At one stage, a little bulge develops on each side of the nervous system, which will become the retina of the eye. Shortly after, this bulge approaches the skin and this develops a little infolding, which will become the lens of the eye. They get together and form a very fine organ. If the retinal bulge is cut away and put somewhere else, then the skin over the new locus develops a lens while that over the original bulge remains skin. Or, at another stage when the vascular system begins to form, most of it develops into blood vessels, but one part forms a heart – that part which is next to the primitive liver, which was formed from the gut a little earlier. If the liver is moved somewhere else, the heart forms there. So these are contingent specifications; and the obvious relation is a spatial situation in the larger unit, in the hierarchically superior entity, in the supersystem.

Incidentally, this could be tied up with some things Tonge said about his searching scheme; whether there is a simple tree and complete delegation or centralization, or whether the operating program simply sends down certain strategies to the subordinate programs, which must operate along those lines. Dr. Mesavoric also made comments that are very relevant to this situation.

Gradients. I would like to question one or two examples of hierarchical relationships that were offered. The Hughlings

Jackson gradient of control in the nervous system was offered by Dr. Whyte as being an hierarchy. Maybe in some respects it is; but it is more clearly a gradient of dominance and subordination at one level. And certainly the suggestion of Dr. Leake, that the nervous system is hierarchically above other organs, and that the gonads (the sex cells) are hierarchically above other cells, is basically not a solid suggestion in my opinion, because these are all clearly at the same level. Organs are at one level and cells at a lower level; what we are dealing with in these cases is not a hierarchy, but a gradient, a gradient of dominance and subordination.

This relationship was clearly recognized by the zoologist, Child; I offer you a clear example from his studies. He showed that the gradient in biological systems was primarily a rate gradient, an intensity of metabolism gradient, a diffusion gradient, or something of this sort; but nothing mysterious. His classical work has done mainly on flatworms. These animals are made of a variety of cells and structure; eyes at the upper part of the nervous system, a pharynx at a lower part, and further down an intestinal tract. There is no question as to which are head structures and which are tail structures. If this animal is cut in two, the head end regenerates a tail and the tail end regenerates a head. So you get two for one by that simple operation, you can multiply by dividing.

Note, however, that the actual position of the cut is arbitrary and a particular group of cells behind it could equally well redifferentiate into head structures, or those in front of it become tail structures. The cells are the same, it is a different relationship to their fellow cells that is important here. Again a kind of global or supersystem variable is operating. If one cuts out a sufficiently narrow piece, especially if it is near the head end, each cut starts to regenerate a head. So clearly, cells which are near the top of the gradient tend to form heads and those near the bottom tend to form tails. We are familiar with this situation in social systems. When the commanding officers of a military unit are lost, the subordinates take over and exercise the command function. The same is true in almost any organization; it just isn't as likely to happen so dramatically. Conversely, when the lowest level elements in an organization are removed, the ones just superordinate take over; low level supervisors in the telephone company man the switchboards when the operators strike.

Integration. Another point concerning interactions is of utmost importance. As one moves from loosely integrated systems to more tightly integrated systems or, I could simply say, as one moves toward greater integration within the system, the forces going vertically from the larger system to the smaller units increase in power relative to the increase in the forces going from the lower to the higher elements.

This is of great interest at the social level. As man develops more and more complex and tightly integrated societies – and certainly we are moving this direction – I do not see how we can avoid moving from anarchy, at one extreme, toward totalitarianism at the other. This gets to the question of the individual versus the state or the community, of the individual's freedom or restriction of freedom, and to the present problem of individual dissent and the breakdown of collective living. How far do you go? Well, I suggest (this is another whole topic, so I merely introduce it without developing it) that as you live in more integrated systems, you inevitably lose some of the freedoms you had originally. People are very unhappy about this, but they forget that they also gain new freedoms that did not previously exist. Mostly, if one does stop to think about the balance, one is not willing to sacrifice the added freedoms for those lost. For example, in a society with laws or sanctions, you are not allowed to do what you want to the body of someone else; you mustn't kill him or beat him up. But, on the other hand, you do have language with which to discuss your differences. Language you would never have had without the higher level system; it is one of those wonderful creations of the committee of the whole. Thus, we find that "goals" emerge at this superordinate hierarchical level.

Purpose. Goals have been referred to several times in this volume, particularly by Dr. Pattee. It is in conformity to the "goals" of the larger unit that the smaller unit is often constrained in its behavior. I remember again Whitehead's statement about electrons running in circles in atoms. This is so beautifully exemplified in biological systems that it is almost impossible to avoid the notion of purpose when examining these mechanisms.

For example: without food many organs waste away, the body burns fats, carbohydrates and proteins to supply energy for two organs which don't waste at all (or not until much later) and which are clearly essential to the continuing functioning of any of the organs – the nervous system and the heart. One can get along without a digestive system quite well under starvation conditions, a great wasting of muscles can be tolerated, the reproductive system isn't important, and so on; but if the heart stops pumping or the brain functioning, the whole system is gone. So there is a sort of value judgment in the way the body responds.

In cold the same "purpose" shows. The body must preserve heat; one of the ways it does this is by constricting the blood flow to the surface of the skin to cut down radiation and conduction and the rest. This may go to such an extent that the skin is frozen and dies. So, here you have the sacrifice, if you will, of the subordinate unit for the protection of the larger unit. And goals in general are adaptive kinds of behavior.

I dare say that this is a broadening of the basic insight of Darwin. Evolution is a matter of the survival of the fittest; natural selection sees to it that those organisms which are more adaptive to their environment somehow survive. Selection is by reproduction and destruction, the birth and death processes. These are all implications of whole and part – the unit, the interactions between units, and the interactions that may exist between any levels. Division of labor with specialization and with progressive reintegration of the specialized units is involved. I have already referred to the degree of integration; in evolution, certainly in the biological domain and perhaps at all levels of material evolution, I see a progressive change towards more and more highly integrated systems. One sees this certainly with the individual and society, and it always has survival value.

LEVELS

I was not convinced by Dr. Bunge that one of the necessary prerequisites of a hierarchical situation is that there be a top boss. There certainly are levels of authority, but *a* boss implies *an* individual element having top authority, and this is not necessarily the case. I had some fun with my Russian colleagues, when I lectured in Moscow, by pointing out that in the nervous system, which is a pretty effective system operating many units in a coordinated and efficient manner, there was no single neuron or small group of neurons that always gave the final orders; but that, in effect, that group of neurons which had the most, and most relevant, information in the immediate situation was the one from which the action started. And that is more like the organization of democracy than of a monolithic totalitarian state. Again, I think such a flexible authority fits some of the points that Tonge made in relation to his computer languages where strategies, not specific orders were passed along.

The evolution of systems is toward increased integration, at whatever level; then new levels are superposed. And as more superordinate levels appear, more and more history is built into the system, as Cyril Smith emphasizes. Multilevel systems cannot form *de novo*; instant humans or instant societies are not possible as is the instant fusion of subnucleons. History becomes progressively important at the higher levels; the uniqueness of each individual representative becomes more dramatic at the higher levels; there are more *kinds* of individuals in a class, such as a species; but there are fewer individuals, as compared with the lower levels.

There is some discussion in this volume about the relation of a system to a hierarchy; and some tendency to distinguish between them. My own thinking leads me to reject this because "system" implies an entity containing subordinate units in some relationship to each other; and that implies hierarchy, superordinate and subordinate levels. Conversely, hierarchy implies units related to each other at different levels; and that is a system. So I find it premature to attempt to separate these two different terms.

Another point. As one looks at the different levels one can see them ordered along the dimension of the *origin* of the units, or along the dimension of the *function* of the units. The morphology of units is likely to be related to origin; so one finds, for example at the biological level, that different appearing cells at one level constitute the tissues and organs at a higher level. Organs are different kinds of cells functioning together. I used to wonder a good deal whether a tissue is a level below an organ or equal to it or above. They are coordinate, I think. The tissue is a group of cells which are identified by their origin or structure, they all come from the same parent cell at one point and all are more or less alike in structure; the organ is made up of cells of different origin and structure, but pulling together in a common function. A like double line of hierarchy exists also at the level of groups of organisms; the species is based on origin, and the ecological group, at a like level, is based on the functional relations between the units – the termites and their cows, for example, function as ecological or functional multiorganism systems. The same sort of thing is very clear in society; the "role" is a functional concept; the individual occupying it is a structural one. Unfortunately, the two are often not well matched in current social ways of casting individuals in roles. In the non-living hierarchies, one line goes through particles, atoms, molecules, crystals, and another goes through geological formations and astronomical systems.

History. Next, a further word on history, where I find this sort of structuring of the situation helpful. One often says that structure determines function and function produces structure, a two-way street, so perhaps the distinction is spurious. I see rather a spiral up the levels, from wherever one starts – let's say history to structure to function to history to structure to function: but each time, going from a lower level to a hierarchically higher one. So movement is not around a circle, but up a corkscrew or spiral.

I have distinguished three basic properties of systems, particularly organisms, which are collectively comprehensive and mutually exclusive. Those characteristics, those anisotropies, if you will, which are essentially constant in time, we call structure or "being" in my alliterative triad; those changes in time which are reversible, which can revert back to the status quo ante, are function or "behaving"; and those changes which are irreversible in time are "becoming." This last subsumes development, evolution, learning, history and such progressive changes from which the system never returns completely to where it was. So history produces structure, which produces function, which produces history, and so on. History at a molecular level, or at least a subcellular level, for example, determines the structure of muscle fibers. These long cylindrical units, extremely regular in their longitudinal (not transverse) internal structure, are made of fibrous molecules all arranged in the same direction. Certainly historical processes laid down this structure and they operated differently in different dimensions. This structure in turn determines the function of the muscle, because only with such an isotropically arranged and correctly oriented structure can a shortening occur with some general change in the system. If structure were homogeneous or randomly heterogeneous, all sorts of tensions might develop, and perhaps squeeze out some fluid, but one end would not come closer to the other while the rest of the muscle bellied out.

Classification. Finally, I shall give you my version of a classification or hierarchy of organizations; not of all possible ones, but of ones we are interested in. I would use "system" for any organized collection of elements with the properties that I have indicated. These include both material and formal non-material systems. I have liked, but can't sell anybody else on it, the word *org* for that subclass of systems which is composed of material systems, in which matter enters into the picture; this excludes formal systems, for example. Then, under orgs one can go along any dimension you want, from the atom to the molecule to the crystal to the various larger units; to geographical, geological, astronomical units, including the various levels of hierarchies in the heavens. This is one line of going up, these are all non-living orgs; in contradistinction to a subcategory of orgs which has the properties of life. This is not a meaningless or tautological statement, because one can define those properties fairly exactly. I would like to call the latter group *animorgs*, which includes all kinds of "organisms." (I don't use the word "organism" in dealing with systems because it has a special biological meaning.) Subclasses of animorgs include unicellulars; multicellulars, the usually described organisms; and epiorganisms, which are not multicellular but multi-individual living systems which include societies such as anthills and villages and the like.

REFERENCES

Gerard, R. W. 1940. "Organism, Society and Science." *Scientific Monthly* 50:340-50; 403-12; 530-35.

——. 1958. "Concepts of Biology." *Behavioral Sci.* 3:92-215. Reprinted, 1958. NAS-NRC Publication 560, pp. 92-215.

——. 1961a. "Quantification in Biology." *Isis* 52:334-52.

——. 1961b. "The Program of the Mental Health Research Institute." *Behavioral Sci.* 6:66-71.

——. 1965. "Memory: Fixation of Experience." In *Science and Human Affairs*, ed. Richard Farson. Palo Alto, Calif.: Science and Behavior Books, Inc.

■

Structure and Function in Living Systems

Herbert Gutman*

Living systems, in their structural and functional organization, follow a hierarchical plan. This note represents a partial excerpt of the argument that an understanding of the genesis of such hierarchy must proceed from a fundamental clarification of the relationship of structure to function and of organic whole to their parts (Gutman 1964). Further, a consideration of the history of living structures is essential in the study of these relationships. This history includes not only ontogeny but the whole chain of evolution.

I have discussed in detail the relationship of structure to function in terms of "existential," "operational," and "organizational" structures (Gutman 1961, 1964). In brief, these differentiate three levels or degrees of freedom of movement of the structural parts. Existential structures are those in which the parts have the least freedom (e.g., spider web, nest of a bird, a bridge, a building, furniture, etc.); in these the relative *positions* are permanently fixed. Operational structures are ones in which some or most of the parts have a limited freedom of movement (e.g., man-made machines and instruments); in these the *pathways* of moving parts are more or less fixed. Organizational structures are those in which most of the parts have relatively large freedom of movement within the boundary of the structure, in some instances even in and out of the boundary (e.g., living organisms on the cellular level and social organizations of animals or man), in these not even the pathways are fixed although the *roles* that these moving parts play are usually well defined. Living systems start out as organizational structures. In their evolution or development, they make increasing use of operational and existential structural elements.

The term function may be used with existential, operational, and organizational structures since the application of the term function is justified when one deals with a meaningful activity or meaningful application: meaningful in terms of a purpose to be fulfilled, an objective to be achieved, or a result to be accomplished. An entity or part has a function if it serves a purpose for man, contributes to the functioning of a whole or constitutes a necessary link in completing a sequence of events that lead to a specific end result. On the level of living systems, I also consider function to be synonymous with role-playing.

The subsequent discussion leads to the primacy of function over structure and points out that fixed structural aspects—be they of rigid

**P.O. Box 356, Topanga, California 90290.*

position or of clearly defined pathways of function—are the results of later "mechanization" of function. These fixed structural properties serve as "memory" of past function. The primacy of function over structure in living organisms is further evidenced by the phenomena of *equifinality,* i.e., the reaching of the same end result from different starting points or different means or routes, and *equipotentiality*, i.e., the fact that very different structures can perform one and the same function. The argument rests primarily on the assumption of a primacy of the whole over the parts in living organisms and that development of parts from a central whole is the result of a progressive individualization and specialization, i.e., a delegation of function to specialized structures. Because of this "top-down" pattern of a process of institutionalization of function, individualization and specialization, we get a hierarchical order.

REFERENCES

Gutman, H. 1961. "The Biological Roots of Creativity." *Genetic Psychology Monographs* 64:417-458.

——. 1964. "Structure and Function." *Genetic Psychology Monographs* 70:3-56.

■

Part IV
Hierarchy in Artifact

No longer is the natural order the sole source of *basic* scientific knowledge. Sophisticated human artifacts such as computers, communication networks, and space systems have joined molecules, stars, and bio-organisms as fruitful objects of study for the discovery of fundamental scientific relations and principles. The study of complex creations of technology frequently produces basic knowledge beyond that used in their design. For example, levels and hierarchies are often designed into man-made systems and organizations, but sometimes they emerge unplanned, as in the discovery that the flow of traffic through the Hudson River tunnels could be increased through the platooning of vehicles.

The examples of important and interesting hierarchical structures in social and technological systems could dominate this volume if they were to be adequately described. But space permits the selection of only two. In the first paper of Part IV, Fred Tonge discusses some of the hierarchies encountered in computer technology and information processing such as; file structures and the organization of computer memories; control hierarchies employing executive programs and user programs with sub-routines and sub-sub-routines. The structure of a program frequently provides an excellent analogue to administrative organization. The forms of general problem solving programs parallel decentralized, centralized, bureaucratic and roving managerial strategies and give insights into the advantages and limitations of each approach.

In the second paper, Robert Lucky discusses the problem of minimizing errors in data transmission codes. The concept of hierarchy enters through the use of concatenated codes or codes within codes to provide cross checks on the accuracy of transmitted data. The mathematical development possible in this subject offers one of the few quantitative approaches to a theory of hierarchical structure available at present.

In the notes that follow, Magoroh Maruyama discusses the levels that occur between the perception of patterns and patterns in social events. Bill Wells comments on the necessity to take into account the level structure of society in the dynamics of social change. In reminding us that the problem of how many parts come to be a unified whole is a 5,000 year old problem, Ronald Jones emphasizes the cultural significance of the study of structural hierarchies.

■

Hierarchical Aspects of Computer Languages

Fred M. Tonge*

Computers, programming languages, and program organization are all manifestations of information processing, and information processing may be a very complex undertaking. This paper is a rather broad look at some hierarchical aspects of information processing in computers. Its basic theme is that hierarchical notions pervade the whole of computer construction, programming, and problem organization.

HIERARCHY

In considering the meaning of hierarchy in information processing, it is useful to turn to the organizational literature. Simon (1960, 1962) holds that hierarchical structure arises in administrative organizations to provide a subdivision of content and a pattern for authority relationships. Analogous to these in information processing would be hierarchical organization of program (function) and data and of the flow of control during processing. Simon argues that such hierarchy arises from a need for stable subunits, of sufficiently little complexity that humans can deal with them, and from a need to minimize requirements for information transfer among such units. Almost the same words can be used in accounting for the pervasiveness of hierarchical structure in computerized information processing.

There are a number of aspects of computation in which we can find hierarchical notions. In the beginning, as it were, von Neumann (1958) introduced a hierarchical organization of computer memories. Most programming languages, whose design reflects both our thinking about programming and the nature of computers, include the concept of a hierarchical organization of subprograms (Knuth 1968; Rosen 1967; and Wegner 1968). And in the dynamic computational process that unfolds as a program operates upon its data to produce results, we also find hierarchy.

The canonical representation of such hierarchies might well be the tree and we shall meet this structure again and again in looking at specific examples of the above aspects of computation.

Information and Computer Sciences, University of California, Irvine, California 92664.

A SIMPLIFIED VIEW OF COMPUTERS

For purposes of discussion, we shall take a simplified view of computers, programming languages, and the dynamic process of program execution.

Hardware. A computer is made up of four types of components:

- *input-output devices,* which are sensors and effectors such as readers, printers, teletypes;

- *memory*, divided into units (words or characters) which hold both the data on which programs operate and the programs themselves;

- *arithmetic-logical unit*, containing adders, accumulators, comparators; and a

- *control unit*, containing such components as instruction decoders, a pointer to the next program step to be carried out, and a pointer to the next piece of data to be processed.

Typically, each program step (or instruction) specifies an item of data and an operation to be performed upon that datum. The dynamic operation of the computer, viewed at this level, consists of a sequence of two-part cycles: fetch an instruction, execute it, fetch an instruction, execute it, ...

Thus, the first of the two pointers mentioned above points to a stream of instructions which flow through the control unit. This notion of a relatively smooth linear stream, as it appears in our simple model, is in fact close to true for most computers.

Narasimhan (1967) points out that "A hierarchical computation can be realized in a simple computer only through the mediation of an . . . executive program or supervisor." Loosely speaking, we should read for "hierarchical computation," one involving closed subroutines and partitionable into subcomputations, for "simple computer," a so-called "von Neumann type machine" (which means most computers currently in use), and for "executive program or supervisor," additional program steps not necessary to the problem solution but to deal with the flow of control among subcomputations. In essence, the statement is that hierarchical programs are not well suited to linear machines. One must add additional (programming) machinery not related to the problem computation, if you will, to achieve the execution of a hierarchical program. And yet, most programs in practice are hierarchically organized, as a practical means of dealing with complexity. Because this machinery requires both processing time and memory space, and because the essence of programming is efficient use of time and space, there is a continuing debate over the manner and extent to which hierarchical notions should be introduced.

Programs. A program to carry out certain manipulations upon data consists of a sequence of instructions and must be expressed in a programming language. Indeed, it is common to speak of a hierarchy of programming languages.

At one extreme of this hierarchy is machine language, in which each statement is a single instruction, encoded so as to make efficient use of the electronic hardware of the computer. An example is the IBM 360 instruction "5A20D00E," which specifies the addition (5A) to accumulator 2 of data in the memory word addressed by 0D00E.

In fact, most so-called "machine language" programming today is done "at the next level up," in assembly language. Here, mnemonic symbols are used in place of the basic machine encoding of instructions. For example, in assembly language the above instruction might be 'ADD 2, FACTOR." Since the computer can only execute machine language instructions, a program, called an assembler, takes as input data a sequence of assembly language instructions and produces as output an equivalent sequence of machine language instructions.

At a higher level yet in the hierarchy are procedural languages, such as FORTRAN, PL/1, or COBOL, which allow the programmer to express his program in a sequence of statements "closer to" the problem-solving procedure he wishes to implement. For example, the FORTRAN statement "X = A+B-D/2" implies a sequence of many machine language instructions. A program called a translator takes as input data a sequence of procedural language statements and produces as output an equivalent sequence of assembly language statements, which an assembler may then process into machine language.

Processes. A process is an instance of a program operating upon data to produce results. It is useful to separate the notions of the program, its data, and the "scratch paper" or working storage needed to keep track of the dynamic application of program to data, a process.

The usefulness of considering processes as separate entities arises in part from the desirability of considering terminating an ongoing process, beginning another, terminating that process to resume the first, and so forth. Why interleave processes so? For example, programs call for input and output, the transfer of data into and out of the memory of the computer. These transfers are generally effected by mechanical devices and so are much slower than electronic speeds. The arithmetic and logical and control units must "idle" waiting for those steps to be completed, and could instead be carrying out another process.

The possible advantage of interleaving processes leads us to the notion of an executive, supervisory, or operating program superior to (that is, hierarchically above in a control sense) the "user programs" discussed so far. Such an executive could decide which programs and data are to be in the computer's main memory, which processes to initiate and suspend, and so forth. Typically, it could also provide a set of standardized services, such as communication with input-output devices. Indeed, today's programmer, unless he is using one of the new small machines now becoming available, programs in terms of a combination of hardware, language translator, and operating system.

TRANSLATION

Because most programs are written in a procedural language or in assembly language, the translation and assembly processes mentioned above form a noticeable share of the load on many computers. An important data structure used in this translation process is the symbol table, as shown in Figure 1. This table provides a correspondence between an external, mnemonic symbol and a machine language datum that can actually be processed by the hardware.

ALPHA	153E
BAKER	0026
SUM	39A1
ZOT	–

Figure 1 Symbol Table

Efficient handling of a symbol table may dictate that it be organized in a less linear fashion, perhaps in a hierarchy as shown in Figure 2. A number of translators so represent symbol tables in practice (Knuth 1968; Rosen 1967; and Wegner 1968).

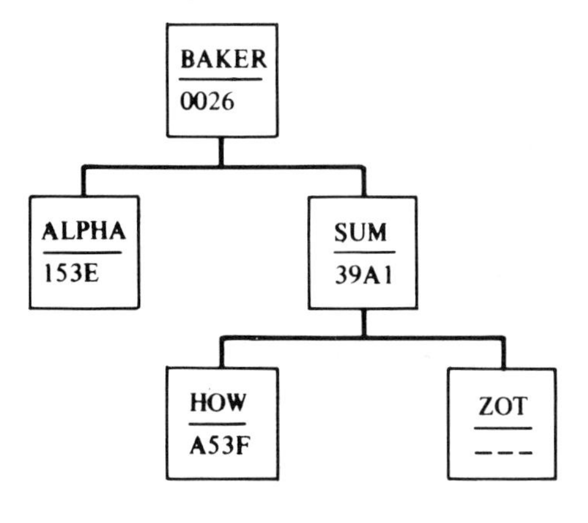

Figure 2 Symbol Tree

Another use of hierarchical data organization in translators is in the representation of the procedural language being translated. Some translators represent that language with a formal syntax, as shown in Figure 3. Input programs are matched with this formal syntax to determine if they are legitimate statements. Often this matching process proceeds by constructing a hierarchical structure representing the levels of formal definition and their final equivalence to the input. Such a

<ASSIGNMENT>	is defined as	<VARIABLE> = <ARITH EX>
<ARITH EX>	is defined as	<TERM>
	or	<ARITH EX> + <TERM>
<TERM>	is defined as	<FACTOR>
	or	<TERM> * <FACTOR>
<FACTOR>	is defined as	<VARIABLE>
	or	<INTEGER>
	or	(<ARITH EX>)

Figure 3 Formal Syntax

structure is given in Figure 4 for part of the statement "X=NU*(Y+15)."

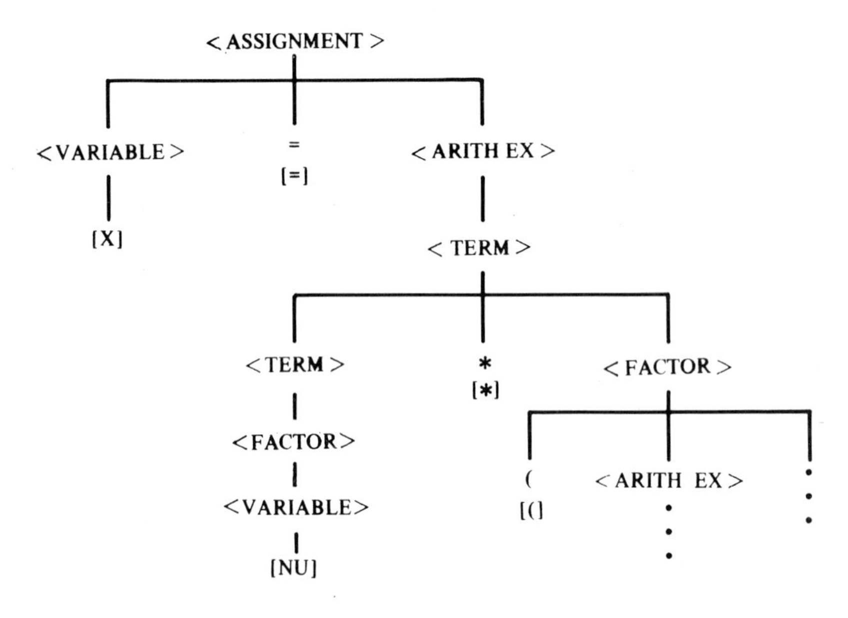

Figure 4 Syntax Tree

NAMING AND ACCESS

One area of programming in which hierarchical notions occur frequently is in the naming of and access to programs and data. Because programming is often a large and complex undertaking, both programs and data are broken up into smaller segments, often handled by different individuals, which must then be joined together to produce the final process. This joining together introduces the possibility of naming conflicts, that is, of using the same name for different pieces of data. Of course, such conflicts can be avoided by employing a sufficiently elaborate bookkeeping system, but such elaborate systems are also susceptible to human failure. And so programmers have developed a number of techniques for shifting the resolution of such conflicts to the translation programs.

Block Structure. One such technique is the use of block structure to organize programs, as found for example in ALGOL or PL/1 (Knuth 1968; Rosen 1967; and Wegner 1968). Figure 5 gives an example of a program organized in block structure, with such structure defining the scope of applicability of data names.

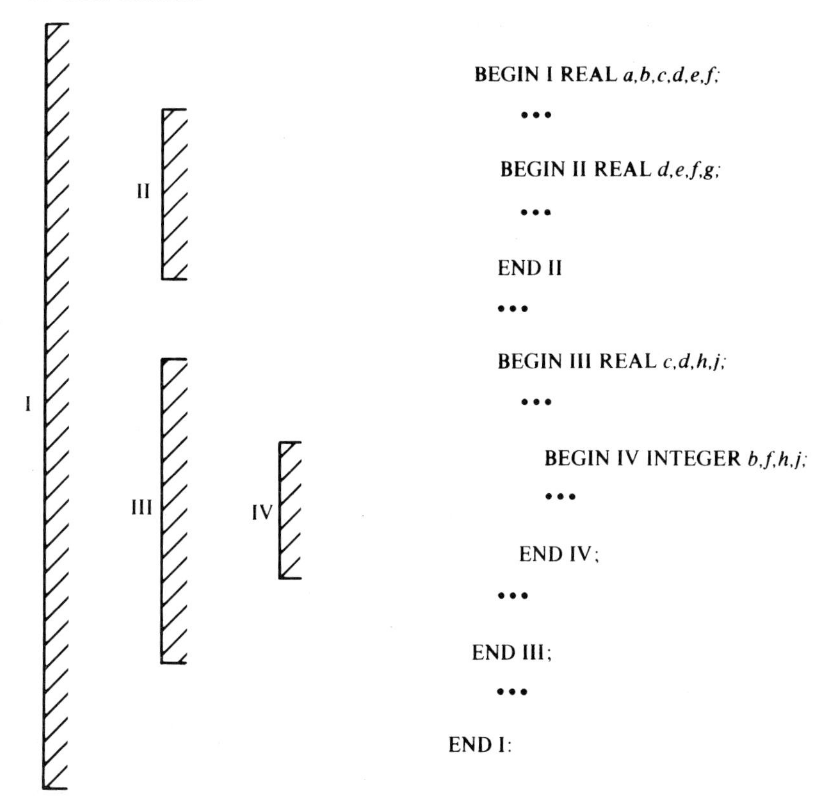

Figure 5 Block Structure

Each block begins with a BEGIN and ends with an END. A datum, such as the variable *a* (whose value here is a real number) is defined in the outer block (I) and in all embedded blocks in which a different *a* is not defined. Thus, the *a* declared in block I is also known in blocks II, III, and IV. But the *d* declared in block I is defined only within I, since both of blocks II and III declare their own, different, *d*'s. Since block IV is embedded in block II, an occurrence of *d* in IV refers to the *d* declared in the outer block III. Such variables are termed "local" to the block in which they are declared and "global" to any embedded block in which they are known.

If we consider, for example, the meaning of *d* to be the program as executed, we see that that name refers to the *d* of block I, then the *d* of block II, then that of block I again, then the *d* of block III (during execution of both blocks III and IV), and finally, upon ending block II, that of block I again.

Record Structure. It is often useful to think of a data record as broken up into fields, which may be broken up into subfields, and so forth. Figure 6 shows an example of two such records, in a manner analogous to how they might be defined in the languages COBOL or PL/1 (Knuth 1968; Rosen 1967; and Wegner 1968). Clearly, this is a hierarchical organization which could be represented as a tree similar to others we have used.

Languages designed for ease in processing such data (as opposed, say, to solving differential equations) typically allow the programmer to define record structures and then proceed using the names such a structure provides. That is, the second subfield of the first subfield of SALES might be referred to as "DAY OF DATE OF

1 SALES	1 PURCHASES
2 DATE	2 DATE
3 MONTH	.
3 DAY	.
3 YEAR	.
2 TRANSACTION	2 TRANSACTION
3 ITEM	.
3 QUANTITY	.
3 BUYER	3 SHIPPER
4 NAME	4 NAME
4 ADDRESS	4 ADDRESS

Figure 6 Record Structures

SALES," or even, since there is no possibility of ambiguity, simply "DAY OF SALES." And "SHIPPER" would refer to both subfields of that field.

We should note that some algebraic languages allow similar naming of groups of fields. For example; in PL/1, the entire 3rd column of a two-dimensional array A(i,j) can be referred to (in some cases) as A(*,3).

File Structure. The organization of data files in the large-scale storage to be used by computer utilities has been the subject of several studies. Such utilities are intended to provide computing and data-handling service to large numbers of customers, and so will have many files of data and programs available for access and modification. Convenient means for referring to a particular file and for getting from one file to another must be provided. In one such project, MULTICS (Daley and Neuman 1965), a joint venture of M.I.T., Bell Telephone Laboratories, and General Electric, files are organized in a tree-like hierarchy. Such a file organization is given in Figure 7.

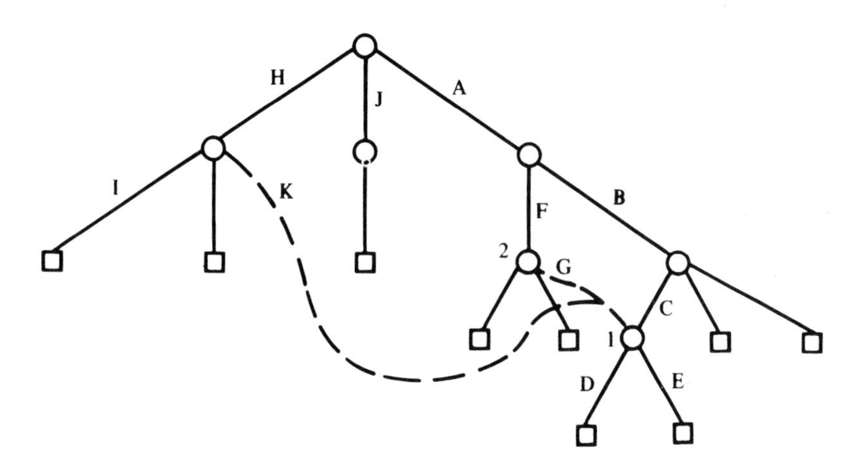

Figure 7 File Structure

The round nodes in Figure 7 are directories to subfiles, which may be further directories or actual data files. Each link is named, and so we may refer to a file by a sequence of links. The actual notation used in MULTICS for naming subfile (directory) 1 would be "A:B:C." To move from one file to another, we use a similar "tree name," using "*" to mean back up a level. Thus, the tree name of file 2 with respect to file 1 would be "*:*:F." Even this use of hierarchy may require long and complicated references, so the capability of defining links which cut across the original tree structure was introduced (dotted lines in Figure 7). We see that file 1 may be referred to, starting from the topmost file, as "A:B:C" or "H:K" or even "A:F:G."

FLOW OF CONTROL

We referred earlier to two pointers within the computer's control unit, one to data and one to the next instruction. The previous discussion concerned references to data, and thus possible contents of the first pointer. Flow of control is concerned with the sequence of machine addresses passing through the second of these pointers.

One simple means of affecting the flow of control which uses hierarchical concepts is that involving the order of interpretation of operations in an arithmetic expression. For example, "A-B·C+D" is not properly interpreted by a straightforward linear scan. It does not mean "subtract B from A, multiply the result by C, and add D to that result," but rather "multiply B by C, subtract the result from A, and add D to that result." Most translators of procedural languages use the concept of a hierarchy of operator precedence (e.g., exponentiation is "more binding" than multiplication or division which are "more binding" than addition or subtraction) to reorder arithmetic expressions so that a linear scan produces the correct evaluation.

The basic hierarchical concept available to programmers to affect the flow of control is the notion of a subroutine. A subroutine is an independent segment of program, written to carry out a certain function, which may be "called" by other

The basic hierarchical concept available to programmers to affect the flow of control is the notion of a subroutine. A subroutine is an independent segment of program, written to carry out a certain function, which may be "called" by other programs when they wish that particular function accomplished. We may think of the calling program as a superior giving a task to a subordinate (the subroutine) who is to report back when the task is completed. Figure 8 demonstrates two uses of subroutine B by "main" routine A. Technically, we may think of the flow of control to the subroutine and back being accomplished in the following manner. The "call" stores in a memory word associated with the subroutine (the linkage word) the address of the next instruction in the "main" routine, and then transfers control to the first instruction of the subroutine. Upon completion of the subroutine, the "return" transfers control to the instruction specified in the linkage word.

Subroutines are widely used and most programming languages provide for their easy implementation. They are used to avoid repeated reprogramming of a common task (as, for example, a subroutine to find the square root of its input) and also as a way of breaking up a complex programming task so that it can be dealt with conceptually be mere people. While the figure does not depict it, we obviously want the possibility of

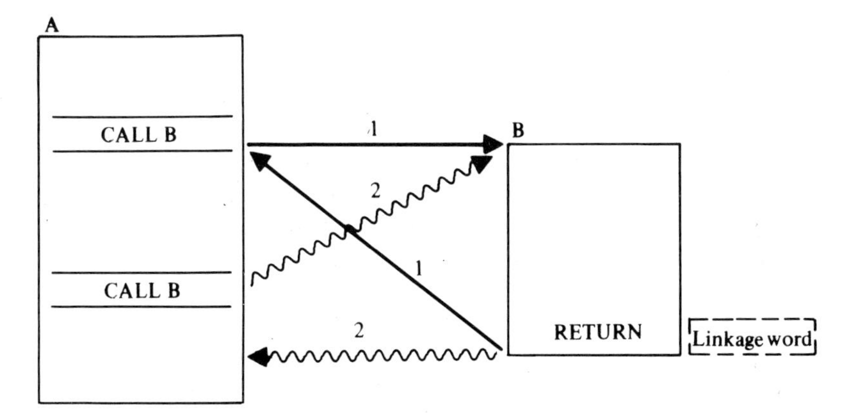

Figure 8 Subroutine

subroutines calling other subroutines which in turn may call other subroutines, and so forth (resulting in a dynamic hierarchy of superior-subordinate relationships as a process unfolds). What may be not so obvious is that we also want the possibility of a subroutine calling itself as a subroutine. The difficulty caused by this possibility, and its solution, are worth exploring.

While there are many examples of subroutines which call themselves (perhaps through a chain of intermediate calls), some of the clearest illustrations arise in dealing with data in tree form. For example, consider searching for a particular symbol in a dictionary tree such as that shown in Figure 2. One definition of the subroutine *treesearch* might be as follows:

If symbol at top is desired symbol, report success.

If not, and there are no subtrees, report failure.

If not, call *treesearch* for the left subtree and, if it succeeds, report success.

If not, call *treesearch* for the right subtree and, if it succeeds, report success.

Otherwise, report failure.

Why won't the simple subroutine linkage scheme given above handle this case? Because, before the first execution of *treesearch* is completed and "return" can make use of the instruction address stored in the linkage word, a subsidiary call to *treesearch* has put a new return address in the linkage word, and so on down through the entire hierarchy of calls. A single linkage word is not sufficient for subroutines that call themselves.

One solution to this problem would be to make a new copy of the subroutine and its linkage word for call, but this could be very space consuming. The typical solution is to replace the linkage word with a linkage stack, as shown in Figure 9. The stack is an area of consecutive words with a pointer to the most recently used word. When a call is issued, the pointer is advanced to the next empty word and the return address stored there. A subroutine return finds the next instruction reference in the main routine in the stack via the pointer, and moves the pointer back one word. A single "central" stack can handle linkages for all subroutine calls, whether the calls are for the same subroutine or another one. (We should note in passing that the same arguments apply to the transfer of input data and output variable names to subroutines.)

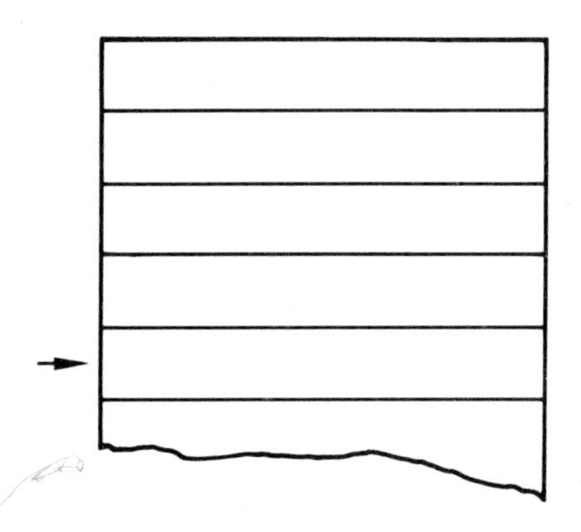

Figure 9 Linkage Stack

FLOW OF INFORMATION

Not only must the program and data be organized in terms of computer processing, but also the flow of problem-solving information must be specified at a higher level. This is particularly true when the problem area is not one in which relatively straightforward algebraic equations specify the computation to be performed, but is rather a highly data-dependent non-numerical procedure, such as proving theorems in symbolic logic or playing chess.

It is typical of such procedures that the problems are combinatorial and are conveniently viewed in terms of the specification and solution of subproblems. This view leads to a tree representation of problem data as it develops, as shown in Figure 10 for a problem in symbolic logic. It is also typical to find a hierarchical organization of the problem-solving procedure.

Allen Newell (1962) has studied the organization of a number of problem-solving programs developed under the rubric of artificial intelligence. He asserts that there is a basic conflict between the hierarchical organization of a problem-solving procedure, which is a means of minimizing the

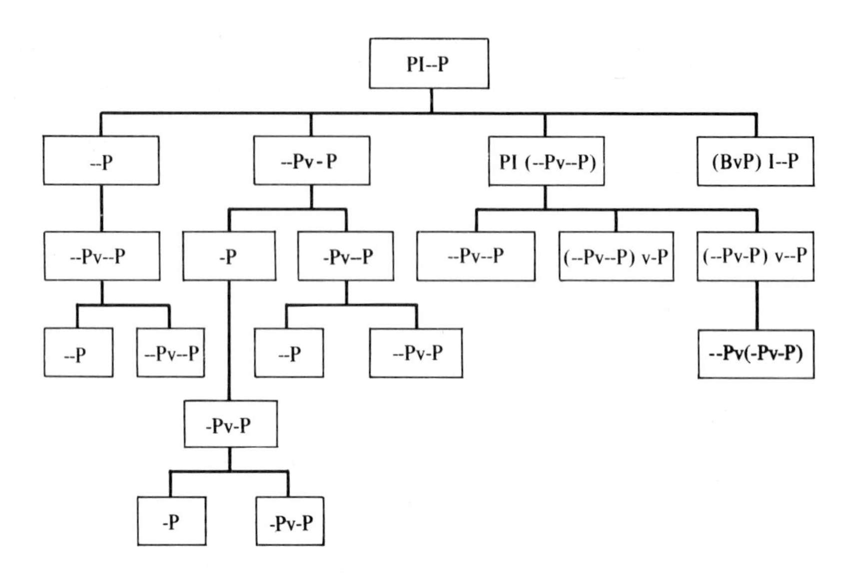

Figure 10 Logic Tree

amount of information transfer required between parts of the procedure, and the richer information environment needed to solve complicated problems. The following comments are based on his discussion of one particular question: What is the problem-solver to do when it has successfully solved a subproblem?

Many early problem-solving programs assumed that subproblems were generated in such a way that their solution guaranteed solution to the total problem. Thus, all subproblems were alike, and need report back only "success" or "failure." For example, in a chess-playing program, a subproblem created by making a legal move would have this property. However, as soon as the problem-solving procedure is generalized to admit different types of subproblems, some of which may not necessarily lead to solution of the "main" problem, a new question is raised. What kind of information is necessary to make use of the results of solving a subproblem?

In a program called the General Problem Solver, Newell and Simon allowed for a number of different types of subproblems. Initially, they solved the question raised above by noting that the superproblem setting up a subproblem (the higher node on the tree, if you will), "knows" what the subproblem is supposed to accomplish and so can properly handle the result of the subproblem given only a report of success or failure. GPS was implemented using this philosophy. However, this approach forces the order in which subproblems are carried out. A superproblem sets up a subproblem, and that subproblem must then be solved so that the superproblem can evaluate the result. And, of course, solving the subproblem may involve setting up and solving sub-subproblems which lead to sub-sub-subproblems, and so forth. That is, this philosophy leads to a depth-first search of the tree of possible subproblems.

Requiring this search strategy is almost as narrow a restriction of the problem-solving process as requiring that all subproblems be alike. It is a complete delegation of control to the subproblem. Other alternative organizations are needed.

One possibility is for the superproblem to activate solution of a subproblem for one step, to say "Go one step (set up on sub-subproblem) and then report back to me; I will decide whether you are to continue, set up a different sub-subproblem, or what." This delegation to one step is in fact centralization; "Don't do anything without my approval." A problem-solving program organized in this manner is quite similar to a centralized administrative organization, and exhibits similar behavior. It displays many of the problems of centralization, in that "the man at the top" is continually called upon to make judgments in situations for which he does not have sufficient detailed information.

Another alternative, which was also implemented in a version of GPS, is for the superproblem to transmit to the subproblem a set of rules stating when to proceed, when to stop, when to come back up the hierarchy for advice, and so forth. That is, the superproblem sends down strategies and the subproblem does a little processing, then examines "the rules" to see if they fit the situation and specify what is to be done. The problem-solving process looks much like a bureaucracy in action, with "pieces of paper" floating throughout the organization prescribing how to behave in all situations. The computer spends most of its time not solving the problem but passing around strategies and matching them to the current situation.

A fourth alternative to decentralization, centralization, and bureaucracy is to view the situation as one in which a single problem-solver "wanders around" the tree of partially solved subproblems, doing a little bit here and a little bit there in an attempt to move toward a solution to the main problem. This approach is very difficult and complex to implement, because it requires that a great deal of information be available to the problem-solver, that an additional network of information be superimposed upon the subproblem tree.

There is no single answer to the question of how complex problem-solving procedures are best organized (nor of how administrative organizations should be, either). But it is clear that hierarchical notions arise not only in terms of programming languages and data, but also in terms of the organization of the overall problem- solving strategy.

IMPACT OF HIERARCHICAL NOTIONS ON COMPUTER HARDWARE

Many of the hierarchical notions mentioned above have been introduced into the hardware of computers. Most modern machines have provision for implementing single level subroutine calls marking the return address with a single instruction. And a few even have mechanized the linkage stack or list.

The so-called "third generation" machines, such as the IBM system 360, have also introduced hardware capabilities relating to the need for an executive or operating system superior to "user" programs. There is a distinction between the computer being in "supervisory" state or in "user" state. Certain "priviledged" machine operations can be executed only in supervisory state. Areas of memory can be protected from access by user programs. And unusual actions such as machine errors or arithmetic overflows cause interrupts in the user program, with control returned to the supervisor for appropriate action.

PARALLEL COMPUTATIONS

There has been a great deal of interest in, and a few instances of, computers organized to carry out computations in parallel. Since such an approach is potentially divergent from the hierarchial notions which pervade most present computer usage, it raises many interesting and unsolved problems of program organization (Newell 1960).

In a summary of parallel systems, Murtha (1966) distinguishes among four types of parallelism. Two of these types are common in many machines today: multiple bit, as in parallel adders in arithmetic units, and multiple function, as in many current systems which have both a "central" processing unit and input-output processors operating in parallel. On the other hand, designs have been proposed for systems in which an array of simple units carry out the same operations in parallel. And there are also proposals for systems with multiple instruction streams (multiple, parallel, flows of control), so that a number of programs (or parts of the same program) are carried out in parallel. Such system organizations are becoming more and more feasible as hardware technology advances, and they raise questions of program organization to which the hierarchical notions developed above may be ill-suited.

One approach to parallel organization of routines, which has been occasionally but not widely used on present machines, is that of the coroutine (Knuth 1968; Rosen 1967; and Wegner 1968). If we view subroutines as program units which start afresh when called and always return control upon completion to the calling routine, we can view coroutines as program units that resume their previous computation when activated and which can release control when necessary to other routines specified when they are activated. We can picture coroutines as shown in Figure 11, with couplers which can be dynamically mated so that control flows among a set of coroutines with no notion of hierarchy. Because for some computations the notion of subroutine hierarchy does not match the problem and leads to excessive amounts of "artificial" bookkeeping, coroutines occasionally have proven to be a powerful method of program organization.

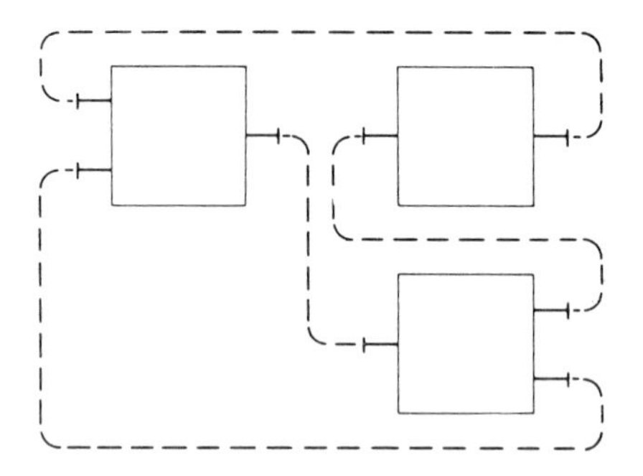

Figure 11 Coroutines

SUMMARY

Hierarchical notions pervade computing. Hierarchy appears in the organization of program units, in the organization of data, and in the organization of information flow during processing.

Hierarchy is dictated in part by the sequential nature of current computers and by our thought processes in attempting to deal with complex information processing tasks. Parallel machines and programs will pose new problems in the organization of computations.

REFERENCES

Daley, R. C., and Neumann, P. G. 1965. "A General-Purpose File System for Secondary Storage." In *AFIPS Conference Proceedings,* Vol. 27, Part 1. 1965 Fall Joint Computer Conference.

Knuth, Donald E. 1968. *The Art of Computer Programming: Volume 1, Fundamental Algorithms.* Reading, Mass.: Addison-Wesley.

McIlroy, M. Douglas. "Coroutines: Semantics in Search of a Syntax." Undated working paper, Oxford University and Bell Telephone Laboratories, Inc.

McKeeman, W. M. 1967. "Language Directed Computer Design." In *AFIPS Conference Proceedings,* Vol 31. 1967 Fall Joint Computer Conference.

Murtha, John C. 1966. "Highly Parallel Information Processing Systems." In *Advances in Computers,* Vol. 7, eds. F. Alt and M. Rubinoff. New York: Academic Press.

Narasimhan, R. 1967. "Programming Languages and Computers: A Unified Metatheory." In *Advances in Computers,* Vol. 8, eds. F. Alt and M. Rubinoff. New York: Academic Press.

Newell, A. 1960. "On Programming a Highly Parallel Màchine To Be an Intelligent Technician." *Proceedings of the Western Joint Computer Conference.*

——. 1962. "Some Problems of Basic Organization in Problem-Solving Programs." In *Self-Organizing Systems,* eds. Yovits, Jacobl, and Goldstein. Washington D. C.: Spartan Books.

Rosen, Saul, ed. 1967. *Programming Systems and Languages.* New York: McGraw-Hill.

Simon, Herbert A. 1960. *The New Science of Management Decision.* New York: Harper & Row.

——. 1962. "The Architecture of Complexity." *Proc. Amer. Philos. Soc.* 106:467-82. New York: Harper & Row.

Von Neumann, John. 1958. *The Computer and the Brain.* New Haven: Yale University Press.

Wegner, Peter. 1968. *Programming Languages, Information Structures, and Machine Organization.* New York: McGraw-Hill.

Weizenbaum, Joseph. 1968. *The FUNARG Problem Explained.* Working paper, Massachusetts Institute of Technology.

■

Structure in Error-Correcting Codes

R. W. Lucky*

This symposium presents a difficult environment for discussion of codes used in the communication of information. Though all of our disciplines tend to get rather specialized, the layman often feels that he has some familiarity with the subject matter. Unfortunately, in the case of coding theory most people tend to think of codes used for the encryption of messages, rather than the type of error-correcting code which I discuss here. For that reason most of this paper will have to be used in describing the subject itself, before I can begin to introduce the notions of structure and hierarchies.

I'm used to this difficulty. At cocktail parties, business executives or lawyers (or those from other occupations who have their own TV shows) often ask me what I do for a living. I used to take the question very seriously and go on at some length to try to explain the intricacies of coding theory, but before I finished the other person would drift away. I found a way to avoid this. Now I simply say that I work for the telephone company.

It is ironic that as an employee of the telephone company, I should be speaking of any hierarchy other than the telephone system itself. The system of telephone interconnections is certainly the most extensive man-made hierarchy. To connect every telephone subscriber to every other subscriber would take on the order of 10^{15} wires if we did not use a hierarchical switching arrangement. Each subscriber is connected to a local switching office, the local offices are in turn connected to regional switching centers, and so on up to five hierarchical levels.

This is a fairly straightforward examlple of hierarchy. The type of structure I discuss here – the structure of codes – is a more subtle and mathematical example. As a reward for trying to understand what I'm sure will be rather unfamiliar material, I can promise some beautiful examples of mathematical structure and some rather intriguing paradoxes. We will see that the

*Bell Telephone Laboratories, Holmdel, New Jersey 07733.

coding problem involves the arrangement of a great many points in some high-dimensionality space. The points can be arranged randomly; that would suit our purposes nicely. However we must remember where all the points are located and there are too many to write down each position. Therefore we turn to simple mathematical rules or recipes for generating the locations. As soon as we use these rules we find that the point locations no longer appear sufficiently random. Can we emulate randomness and yet retain a simply described mathematical structure? The answer appears to be no, but we will see where a sort of hierarchical structure helps.

All of this is getting ahead of the story. Let's start back at the beginning. More and more these past few years we have needed to provide communication between machines as well as humans. These machines, perhaps computers, speak in a simple binary language. They communicate with 1's and 0's. In the ordinary handling of these messages, over radio links, telephone lines, or satellites, unwanted disturbances tend to corrupt or distort the messages in random ways. In a telephone conversation, we often hear snaps, pops and hissing, but we can reject this noise as meaningless. Unfortunately these same phenomena often cause the mechanized system to mistake a 1 for an 0 or vice versa.

ERRORS IN TRANSMISSION OF DATA

The computer accepts these errors in a very literal way. A message such as 10110 · · · has as much meaning to it as the incorrect message 10100 · · · . A spoken or written language on the other hand, incorporates so much redundancy that analogous errors are usually automatically corrected by the recipient. For example, if I went into a grocery store and asked for a loaf of *tread,* the clerk would undoubtedly give me *bread* without further question. Nothing else would make sense.

We need to incorporate the same kind of forgiving redundancy in the computer language in order to allow for the inevitable errors which occur in transmission. This redundancy is added in the form of *check bits* – extra digits appended to each block or word of digits transmitted. For example, a single bit added to each block indicates whether the block contains an even or an odd number of 1's. The block 1010 has an even number of 1's. To indicate this we add a check 0 and send 10100. Now if the second bit is in error at the receiver, 11100 is received and the receiver can tell immediately that the block contains an error since the check bit doesn't agree with the oddness of the block. Although the receiver knows that an error has been committed, there is not enough information in the single check bit to enable the receiver to correct the error which occurred. Usually in simple systems of this sort, the receiver requests the transmitter to repeat the word in question.

We shall not be concerned here about these detection-retransmission systems, but rather with systems in which the receiver has enough information in the form of redundant bits to make an intelligent guess as to the correct word just as the grocer could guess that *tread* should be *bread*. As a crude example, we could imagine that each bit were repeated three times, the two repetitions being the redundant portion. A 1 becomes 111. If the second bit were received in error, we could guess that the resulting 101 had originated from a 111 and the data bit was consequently a 1. But the 101 could also have come from a 000, with errors in the first and third positions. In this case the data bit would have been an 0. We reason that a single error is more likely than two errors in the same three-bit sequence; therefore on receipt of the 101, we declare that a 1 was transmitted.

Because this code can be fooled by two or three errors in the block of three bits, there is still some probability that errors will be accepted. However, the more likely event is that a single error is corrected and the coded system has a lower probability of error than an uncoded system. Notice, though, that the penalty paid for this lowering of probability of error is a decrease in the rate of information transmission. Only one third of the bits actually carry information, the other two thirds are redundant. The code rate is said to be one third.

All this seems obvious. If we want the probability of error to be arbitrarily small, we could certainly repeat each bit a large number of times. But then the code rate would approach zero.

It seems that if we want perfect accuracy, we must settle for zero rate. It was Claude Shannon's great discovery in 1948 that there exist coding schemes for which the probability of error could be made arbitrarily small *without having the code rate approach zero*. This perfect accuracy could in fact be obtained at any code rate less than a certain critical rate, called channel capacity (which of course depends on the probability that any individual bit is received in error). Later I will indicate heuristically how this can be.

THE CODING PROBLEM

Now let's take a more general view of the coding problem. In Figure 1 an error control system is shown with the incoming data bits blocked into groups of k bits (the example shows 4-bit blocks). To these k bits are appended additional, redundant bits which in some way depend upon the k information bits. The total block size to be transmitted is n bits. In the specific example indicated three check bits are added to each four-bit sequence to make a block size of seven.

We should think of this operation of encoding as a mapping. There are 2^k possible input sequences. To each of these possible input sequences we must assign an n-bit coded sequence. Since n is larger than k, the total number of possible n-bit sequences, 2^n, is larger than the total number of sequences used as inputs. We realize that the necessary redundancy arises from this restriction of the allowable output sequences to a subset of the set of possible combinations, just as mpeq is not a word – not all combinations of letters are considered intelligible words.

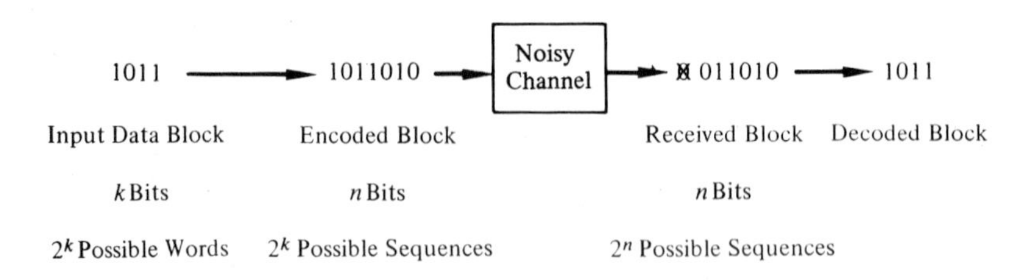

Figure 1 Error Control System

(Incidentally, languages differ in the amounts of redundancy they incorporate and the ability to construct crossword puzzles is indicative of how little redundancy is present. In Hebrew it is possible to make certain simple three-dimensional crossword puzzles.)

In Figure 1 there are $2^4 = 16$ possible input words. With seven-bit output sequences there are $2^7 = 128$ possibilities from which we need to assign only 16 as being "meaningful" combinations. Due to errors any of the 128 possible 7-bit sequences may appear at the input to the decoder. The decoder must then map the particular received sequence back into what it considers to be the most likely of the 16 allowable words.

Conceptually, the encoding and decoding operations are done by dictionary lookup. The encoding dictionary for the example of Figure 1 would have 16 entries, corresponding to the sixteen 4-bit words. Under each entry would be listed the particular 7-bit sequence to be transmitted. The decoding dictionary would have 128 entries corresponding to all possible received 7-bit sequences. The dictionary would list the preferred 4-bit word to be decoded for each of these 128 entries.

Since this dictionary size is too large to use for illustrative purposed, I have chosen to show the encoding and decoding dictionaries for a very short (3,2) code in Figure 2. The notation (3,2) means that there are two information bits in each block of size 3. Notice that the process of encoding and decoding can correct some transmission errors, but by no means all possible error patterns.

Encoding Dictionary	Decoding Dictionary
00 → 000	000 → 00
01 → 011	001 → 01
10 → 100	010 → 00
11 → 111	011 → 01
	100 → 10
	101 → 11
	110 → 10
	111 → 11

Figure 2 A Simple (3,2) Code

It is helpful to look at a geometric picture of this code which is shown in Figure 3. The 3 bits of the code are indicated along each of the 3 axes. The 8 possible combinations are the corners of the cube. Of these $2^3 = 8$ combinations only $2^2 = 4$ are to be used as valid code points. The corners marked by heavy dots are the code points as given by the encoding dictionary.

At the receiver any of the 8 corner points could be received. The decoding procedure must move each corner point back to one of the heavy dots. For example if 101 is received, it could have originated from either of the valid code points 111 or 100 and the presence of a single error. It could also have originated from either 000 or 011, but either of these possibilities would imply the presence of two errors. The decoding dictionary of Figure 2 dictates that 111 is to be used in the decoding of 101.

The geometrical analog in Figure 3 is that errors correspond to distance moved away from a given code point. For each error we move one unit along a particular axis. Obviously the further away the code points can be kept from each other, the more protected we are from the perturbing influence of errors. The fewer code points there are in a given size space, the further apart they can be kept. But of course if there are fewer code points, then there is less information carried by a word – the code rate is lowered.

In coding there are two related problems – compiling the coding and decoding dictionaries. Alternatively we can think of

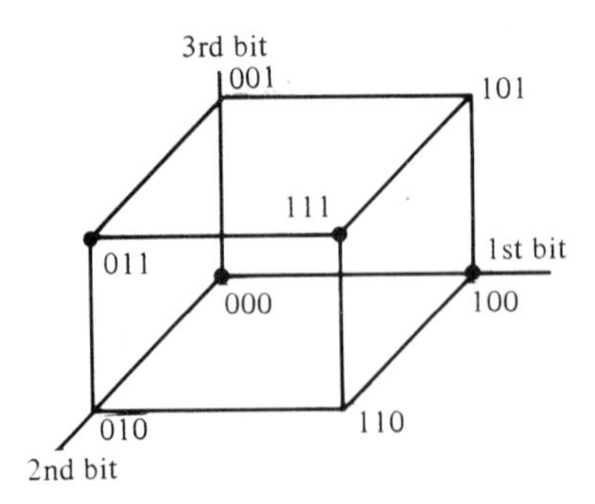

Figure 3 Geometric Picture of the (3,2) Code

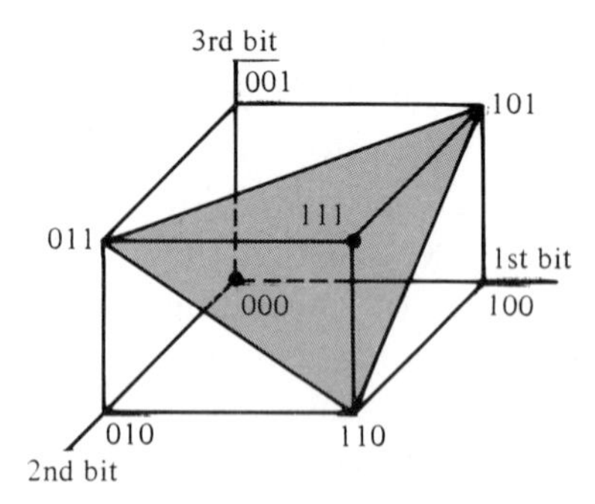

Figure 4 Geometric Picture of the (3,1) Code

the coding problem as the choosing of an appropriate subset of the allowable points in a given space and the decoding problem as the *partitioning* of this space into regions into which we designate a given word was the one most likely transmitted. For a given size code we would like to choose these dictionaries so as to minimize the probability of being in error after the decoding operation. Let's look first at the decoding problem. There is an optimum decision rule, and it's a very simple one. To decode a given received word we simply choose the *closest allowable* code word, i.e., the code word which differs from the received word in the fewest number of positions.

This minimum distance decoding rule leads to the smallest probability of error after decoding (provided that the errors on the channel are independent events, which we always assume here). It also tells us how to optimally partition the code space into "decision regions." In Figure 3 these regions are a little ambiguous because some points are at equal distance from two different allowable code words. In Figure 4 the same space is shown, but only two code points are designated so as to make a (3,1) code. The minimum distance rule tells us that all points on or above the shaded plane are to be decoded as 111, while all points below are said to originate from 000. The probability of making a mistake is the probability that 000 is transmitted and the received sequence lies *above* the shaded plane, or that 111 is transmitted and the received sequence lies beneath the plane. We could use elementary probability theory to evaluate this probability of error.

Now we have a means of evaluating a given code. An optimum decision rule is known and we can in theory calculate the final probability of mistake. It remains to choose the code (coding dictionary) so as to minimize this mistake probability. Conceptually this seems straightforward. Certainly in Figure 4 choosing the code points as opposing points on the cube is best. The idea, of course, is to get the code points as far apart as possible.

The realities of the problem put this simple picture to shame. Practically speaking codes have lengths of, say, between 20 and 100. Theoretically we desire to evaluate the probability of error for very long codes – codes with lengths of tens of thousand and more. Say the code is a (63,51) code. Then we are working with 2^{51} points in a 63-dimensional space. Evaluating the probability of error, let alone optimizing code point placement, seems out of the question.

RANDOM CODES

Shannon brought a very powerful and clever technique to bear on this seemingly-hopeless problem. He conceived the idea of a *random code.* Suppose that the code dictionary is chosen entirely at random. For the (63,51) code each of the 2^{51} entries in the coding dictionary could be chosen by flipping a coin 63 times to get a 63-bit word which we would write down for a particular code point. After 63 x 2^{51} flips the dictionary would be completed. Now an optimally chosen code must be at least as good as this randomly-chosen code. Therefore, if we could evaluate the probability of error for the random code, we would have an upper bound for the probability of error for the best code. Of course if we can't evaluate a given code, how can we evaluate the average probability of error for an ensemble of randomly-chosen codes? Amazingly, this latter average can be found.

By this random-coding argument Shannon found that the probability of error of an optimally-chosen code of length n with rate $R = k/n$ was bounded by

$$P\,(\text{error}) \leqslant e^{-nE(R)} \tag{1}$$

where $E(R)$ does not depend upon the code length n, and is positive for all rates R less than channel capacity C. A sketch of $E(R)$ is shown in Figure 5.

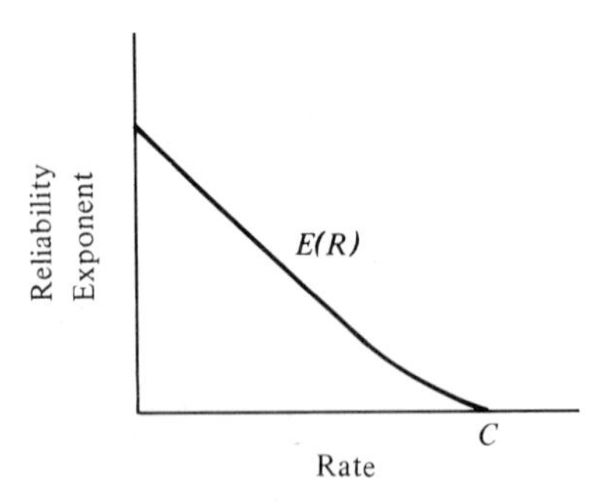

Figure 5 Reliability Exponent for Shannon Coding Theorem

What this bound tells us is that there exist codes for which the probability of error can be made exponentially smaller by increasing the code length, *while maintaining a constant code rate*, e.g., 2/3 of the bits are data bits and 1/3 are redundant. This can be achieved so long as the rate we choose is less than capacity. The nearer we try to approach capacity, the smaller is the exponent $E(R)$. Consequently, to maintain a given probability of error the code must be proportionately longer.

In order to make this behavior quite clear, let's look at Figure 6 where a sequence of codes of rate 2/3 is shown with sample segments of coding and decoding dictionaries. The first code is the (3,2) code used as an earlier example. A (6,4) code is next; this code has the same proportion of redundancy, but the blocks are longer. Finally we illustrate a (12,8) code, which again has the same rate but a longer block length. Shannon's theorem tells us that if the coding and decoding dictionaries are properly chosen, the probability of error is decreasing as we go to the longer block lengths of the example (6,4) and (12,8) codes. (Properly speaking, the bound is an asymptotic one for long block length and there is some difficulty in its application to these short lengths.)

Coding Dictionary	Decoding Dictionary
(3,2) Code	
01 → 011	001, 011 } → 01
(2^2 = 4 entries)	(2^3 = 8 entries)
(6,4) code	
0110 → 011010	011010, 111010, 010010, 011011 } → 0110
(2^4 = 16 entries)	(2^6 = 64 entries)
(12,8) code	
01101011 → 011010110101	16 different words } → 01101011
(2^8 = 256 entries)	(2^{12} = 4096 entries)

Figure 6 Sequence of Codes With Rate 2/3

Why is it possible to make the probability of error decrease for longer and longer codes while maintaining a constant rate? We should have an intuitive feeling for this from the familiar law of large numbers. If I flip a coin four times we expect two heads and two tails, but we would hardly be surprised if we got only 25% heads (one of four). On on the other hand if we flipped the coin 100,000 times and got only 25% heads we would firmly believe that something was unfair about the coin.

What happens in the case of long codes is that the error patterns we encounter become, with high probability, "typical." In the shorter code lengths untypical things (like the 25% heads) occur more frequently and lead to error in decoding. The coding and decoding dictionaries are constructed so as to protect against these "typical" error patterns. In the limit nothing but these typical patterns occur and the probability of error approaches zero.

Now we must notice a very important factor. The probability of error can be made smaller by increasing the code length. But the size of the coding and decoding dictionaries grows *exponentially* with the code length. The (3,2) code has only 4 entries in the coding dictionary, the (6,4) has 16 and the (12,8) has 256. A practical (63,51) code would have 2^{51} ; roughly speaking this is 10^{15} , a fantastically large number, and a code of this length is not nearly long enough to be effective. How are we to construct and remember dictionaries of such sizes?

We will come back to this central problem a little later. In the meantime let's consider the significance of the random-coding argument. If codes picked at random exhibit on the average such desirable performance, what is the performance of carefully optimized codes? In order to get a lower bound on the obtainable probability of error Shannon invoked the argument that the probability of error for the best code was asymptotically lower-bounded by the same expression as was found for the upper bound by random coding. Let's consider that a moment. Codes picked entirely at random perform on the average as well as is possible. Roughly speaking, this means that all codes are good. We can hardly miss.

We saw a moment ago the practical impossibility of constructing dictionaries of the required size. Therefore we cannot simply flip coins and construct a random code. The only recourse is to replace the dictionaries with algorithms. We need simple recipes to find the proper code word for a given input and to decode from a given channel output. Random codes are useless for this purpose. The words have no simple generating rule; the code has no inherent structure. In the last 20 years mathematicians and engineers have discovered many classes of codes which can be generated and decoded by simple rules. In order to be described so simply, the myriad of code points in such high-dimensional space must have a great deal of mathematically-imposed structure. None of these codes which have been found have the behavior predicted by Shannon's bound. All of them become useless as thc code length is increased. Either the probability of error approaches 1/2 (the data are meaningless) or the rate approaches zero. The paradox is this; the random coding bound shows that essentially all codes are good, but every code which has ever been found is bad! Somehow randomness is a desired property to distribute points more or less uniformly in space. As soon as we impose mathematical structure we lose this essential random character.

Let me now summarize the coding and decoding problems and say a few words about how mathematical structure has been introduced to combat the overpoweringly large numbers involved.

Coding Problem Locate 2^k points in an n-dimensional space (with binary coordinate values) such that these points are "maximally separated."

Decoding Problem Given any point in this n-dimensional space find the closest of the 2^k allowable code points.

Let me reiterate that it is impossible to record all the point locations and that it is impossible to search through all allowable 2^k points at the receiver in order to find the closest code point.

MATHEMATICAL STRUCTURE OF CODES

The kind of structure that codes have is algebraic. Essentially all codes are *group* codes, i.e., the code words form a group. Recall that a group is a collection of elements closed under an operation such as addition, possessing an identity element, and including an inverse of every element. This means that if any two code words are added, modulo-two, another legitimate code word results. For example if 0001011 and 0010110 are code words, then so must be 0011101.

Another structural feature nearly always used is to make the code *cyclic*. In a cyclic code any end-around shift of a word becomes another word. Thus if 0011101 is a word, then so is 1001110, etc. Group codes with this property can be described very simply and compactly. We represent each code word by a polynomial with binary coefficients corresponding to the bits of the word. Thus 1001110 is represented by $x^6+x^3+x^2+x$, while 0011101 is $x^4+x^3+x^2+1$, etc. Then every polynomial in the code must be a multiple of some generator polynomial, which must itself divide x^n+1. In the (7,4) code we have been using as an example in these last few paragraphs, each word is a multiple of the generator polynomial x^3+x+1, which itself divides x^7+1.

This gives us a very compact code description. We need only remember the generator polynomial x^3+x+1. On demand any word of the code may be found. Because of the strong mathematical structure we can use the tools of algebra to determine code properties. It has been shown that a very important code property, called minimum distance, is simply related to the roots of this generator polynomial. The minimum distance is the minimum number of positions in which two code words differ. For our particular example, this distance is three. Since each error moves a word a distance of one, a minimum distance of three guarantees that all single errors can be corrected by proper decoding. The correct code word will always be closest to the received code point if there is one error.

This seems to be a simple and elegant solution to the coding problem. This mathematical structure also gives us an algorithmic means of decoding. In order to be a legitimate code word, the check bits must satisfy certain equations whose coefficients are determined by the information bits. The method of decoding involves changing as few bits as possible, both information and check bits, in order to make these equations satisfied. In code space the decoding procedure can be envisioned as surrounding each code point with a sphere of, say, t errors radius. Any time a received code word lies within a sphere we decode to the center of the sphere, which is the only valid code point in the sphere. We make the radius t as large as possible without having the spheres intersect. Notice that this procedure does not completely fill the code space. This kind of decoding is equivalent to true minimum distance decoding only when t or fewer errors are committed. The computation effort required in decoding increases roughly as the square of the code length n. Remember that in minimum distance decoding we would have to compare a received word with all possible transmitted words to find the closest; this would require an effort which increased exponentially with n.

There are a number of classes of these algebraic codes known. The best and most useful of these classes is the Bose-Chaudhuri-Hocquenghem (BCH) code in which the roots of the generator polynomial are chosen in a certain manner in order to ensure a particular minimum distance between code words. Unfortunately, for the BCH codes as well as other similar mathematically constructed codes, their ability to correct errors does not increase in proportion to their length. If errors in the channel occur randomly with probability p, there are on the average np errors in a block of length n. If the error-correcting ability of the code (t) can keep up with this growth as its length n is increased, then the probability of error can be made arbitrarily small as in Shannon's prediction. Since in the BCH codes this does not happen, they become asymptotically useless for large n. This does not mean than the codes are not practically useful, since for moderate lengths their properties are close to known optimal values. The inherent problem is more philosophical than practical.

Before leaving this phase of our exposition, I want to point out that there does exist a coding and decoding technique which achieves some of what Shannon's results promise. The encoding method is not the block-like structure to which we have thus far referred, but rather a recurrent (convolutional) structure where the check bits are interlaced with the information bits. The check bits are computed by a "sliding window" which uses a certain subset of the information bits within its aperture at a given moment. The result of the encoding procedure is a code which can be described in a tree form. Each time a 1 or 0 is transmitted this gives us a direction to move in the code tree.

A decoding technique known as *sequential* decoding is used. The object of the receiver is, in spite of the presence of errors, to duplicate the path through the code tree used by the transmitter. The decoding algorithm is really a computer search routine, which emulates the behavior of a motorist following a vaguely familiar route. As long as some familiar signposts are seen we assume we're on the right path. If we make a wrong turn, things get more unfamiliar as we proceed. Finally we decide to back up and reverse one of our earlier turns. Is the new path more familiar?

It can be seen that the decoding computational effort is a random variable. If there are a large number of errors, more and more wrong turns are made and things begin to get out of hand. It has been shown that this system can be used successfully to achieve arbitrarily low error probability, as we desire, but only below a certain rate which is less than Shannon's channel capacity. Above this rate the search process fails to converge to the true path. Therefore, although a sizeable inroad has been made, Shannon's capacity remains unconquered.

So far we've had something to say about *structure* in codes. It appears that we want randomness, but because of the exponential growth of dictionary size we can't go about coding and decoding in any other but an algorithmic, structured fashion. Our attempts at emulating random distributions with mathematically contrived arrangements appear to have failed.

Now I want to move in the direction of hierarchically structured codes. The first step in this direction is the *iterated* code, although this is not a structure which could be termed hierarchic.

HIERARCHICALLY STRUCTURED CODES

Iterated codes have been known since the early days of coding, and are in fact used in most magnetic tape recording for computers. An example is shown in Figure 7. Each row of the illustrated matrix constitutes a code word with a single parity check on the right indicating the evenness (0) or oddness (1) of the row. Six of these words are stored in succession at the transmitter. Then a seventh word is computed whose bits are the vertical parity checks on the columns of these words. Thus we have both horizontal and vertical parity checks.

Figure 8 shows how this helps in decoding. Suppose that an error occurs in the bit marked with an X. If we had only the horizontal parity checks, we would note an error in the 4th word, although we would not know which bit was in error. Because of the vertical checks we can further locate the error to the 4th column. Of course this completely specifies the error location and the offending bit can be changed. Notice that the efficiency, or rate, of the initial code is multiplied by a factor of

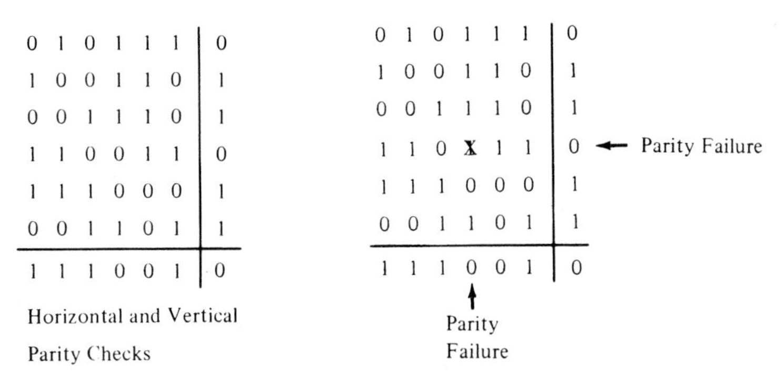

Figure 7 Iterated Codes

Figure 8 Error Correction by Horizontal and Vertical Parity Checks

6/7 because of the further coding in the vertical direction. The code rate in this example is 36/49.

A more general form of iterated code was investigated by Elias in 1954. The iteration can be extended to other dimensions as shown in Figure 9. Here we start with an (8,4) code in the horizontal direction. We save 11 of these words and then append 5 words which act as vertical checks to form a (16,11) code on each column. Now we save 26 of these matrices, append 6 as checks in the 3rd dimension, etc. The sequence of codes used in higher dimensions is indicated on the figure.

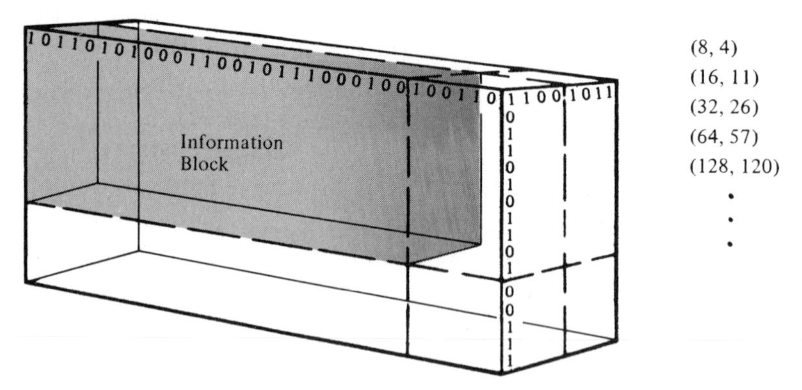

Figure 9 Elias' Iterated Coding Scheme

The code rate in this example is diminished by each further iteration. The information bits after the first two iterations are shown in the shaded solid; all other bits are check bits computed according to certain equations. Observe that the sequence of codes used increases in efficiency, so that further iterations diminish the overall code rate, but by an ever-decreasing factor. It can be shown that in the limit as we iterate an infinite number of times, the code rate appraoches some positive (non-zero) value. At the same time the probability of error approaches zero (if the initial bit-error probability is below a certain threshold). Again we have achieved some of what Shannon predicted. Although the probability of error can be made arbitrarily small, the data rate achieved is far less than the channel capacity obtained by Shannon.

Elias' iterated coding scheme was the first constructive procedure for obtaining arbitrarily small error probabilities at non-zero rates, but for ten years the idea lay dormant. Then in 1965 three persons, Forney in the U.S.A., Ziv in Israel, and Pinsker in the U.S.S.R., announced procedures capable of approaching channel capacity without exponentially growing complexity. Each of these procedures, although differing considerably in detail, was based on the *concatenation* (putting codes within codes) of known codes. In the last few years some of this work has been furthered, notably by Falconer who proposed yet another concatenation scheme. At the moment it is true that all known schemes for realizing channel capacity without exponential growth of complexity involve concatenation. This concatenation is a kind of hierarchical coding structure. It is not known if this kind of structure is essential in realizing channel capacity, but it seems more than coincidental that independent workers in widely scattered portions of the world should simultaneously arrive at such conceptually similar systems.

The kind of code concatenation proposed by Forney (the only procedure we discuss here) is shown in Figure 10. Let's start on the inside of this system and work outwards with the description. The channel transmits digits one by one. We group these digits together to form words in which redundancy is incorporated. This is the *inner code*; in the example shown, the inner code is a (5,3) code consisting of two redundant digits added to each 3-bit block. The input to such a coder is a three-bit block of information, 101, 111, etc. We can in turn consider these *blocks* as *characters* in another code, just as binary digits were the characters in the inner code. This time the characters, instead of being binary numbers, are numbers chosen from a larger field. In this example the characters in the outer coder are chosen from a field of eight elements – the number of possible three-bit sequences.

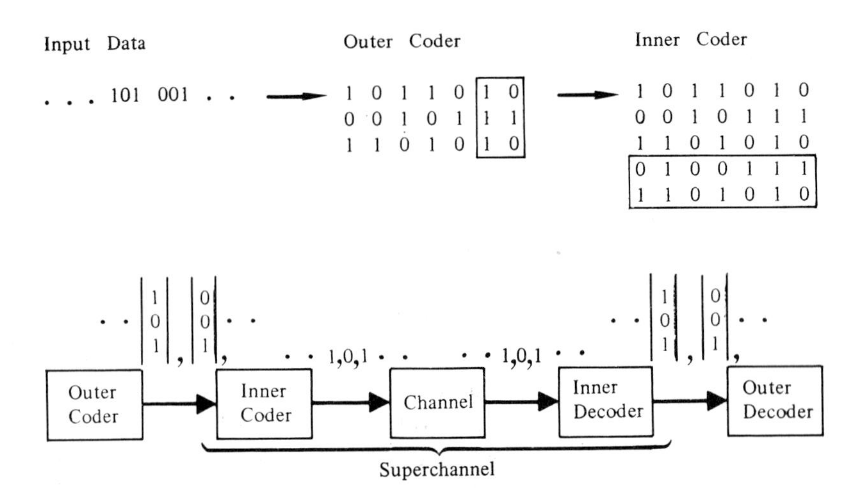

Figure 10 Concatenated Codes

The outer coder groups these larger characters together and adds redundancy in the form of additional characters. In Figure 10 the outer coder takes 5 of the three-bit sequences, which act as its characters, and adds two additional three-bit sequences. The two additional three-bit sequences act as checks; that is, they are computed in some fashion from the five initially specified characters.

This description is probably more complicated than necessary, but the idea is important. The channel deals in individual *digits*; the inner coder assembles and operates on these digits to form *words*; the outer coder in turn uses these words, assembling and operating on the words to form what we will call *superblocks*. The process could be continued. For example there are 8^5 possible superblock inputs to the outer coder. Another coder could be added which dealt in characters which have 8^5 possible values.

The concatenated coding system is somewhat analogous to the way in which redundancy is incorporated in language. The basic building blocks are letters. A group of letters forms a word, but of course not all combinations are valid words. A further grouping of words forms a sentence. Again, we cannot just string words together arbitrarily and obtain a meaningful sentence. The rules for forming sentences from words build in a redundancy, just as the rules for forming superblocks from code words do in the case of digital transmission. Thus concatenated codes parallel the hierarchical structure of language.

In decoding the concatenated code an inner decoder operates on the code words, reducing them to best guesses of their information content. In Figure 10 the decoder accepts blocks of 5 bits and decodes to three-bit words. The outer decoder accepts blocks of these inner decoded words and decodes the resultant superblock to a best guess of its information content. In the illustration the outer decoder waits for 7 of the inner blocks to be decoded and then decodes the resultant 7-character superblock to a 5-character superblock (a 5 x 3 matrix of binary numbers).

One of the exciting aspects of coding theory relevant to the study of hierarchies is that the performance of concatenated codes can be evaluated mathematically. There is a certain job to be done – that of constructing encoding and decoding dictionaries. This job can be done using one huge dictionary in the straightforward manner described earlier, or it can be done using two or more smaller dictionaries hierarchically organized. What do we lose by incorporating such a constraint on our coding method? In Figure 10, for example, the superblock after inner coding is in the form of a 7 x 5 matrix of bits, 15 of these bits are information bits and 20 are redundant. We could imagine an optimum code being constructed which has a block length of 35 and contains 20 information bits. This code must perform at least as well as our concatenated code, since it does not have the constraint of having to be constructed in the hierarchical manner. How much better can this optimum code be made?

By an application of Shannon's random coding bound, Forney has proved a bound for concatenated codes. This bound, shown in Figure 11, is of the same form as the Shannon bound. Using concatenated codes the probability of error can be made exponentially smaller by increasing the overall code length n, while maintaining a constant code rate (fixed ratio of redundancy). In the concatenated code the length n must be considered as the total number of bits (35 in Figure 10) in the overall code matrix.

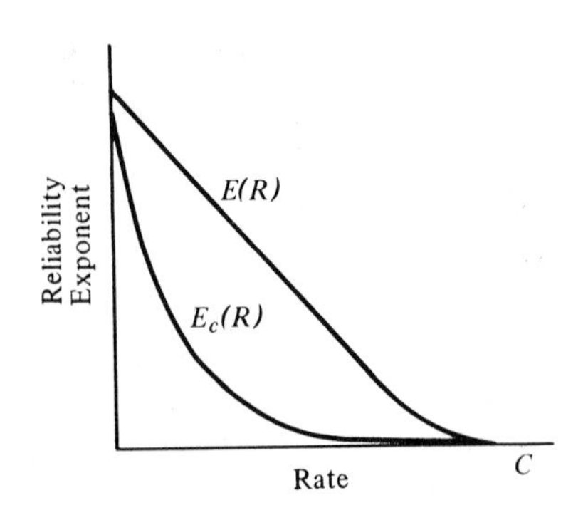

Figure 11 Reliability Exponent for Concatenated Coding Theorem

This behavior can be achieved at any rate up to channel capacity. However the constant in the exponent of the concatenated coding bound is considerably smaller than in Shannon's original bound. This means that, while ideal behavior is theoretically achievable, for a given probability of error a code must be much longer if it is to be formed by the concatenation of two codes than if it can be constructed in one fell swoop. We can estimate from the bound the exact penalty we must pay for having to hierarchically organize our coding system.

It is important to realize that concatenation does involve a penalty, but that ideal behavior is still asymptotically possible. The question is whether or not concatenation makes it possible to construct very long codes without resorting to huge dictionaries. We shall see later that it does, but for the moment we might ask why there is a sizeable penalty associated with concatenation. It seems that requiring a given code length to be the composite of two smaller codes cannot be too severe a structural constraint. Indeed it does not appear that the loss in performance can be attributed so much to the code construction as to the decoding procedure.

In ideal decoding of the example code shown in Figure 10, we should search through all possible valid 7 x 5 matrices to find that which corresponds closest to the received superblock. However the decoding is done in steps. First the columns are decoded into "most likely" 3-bit blocks, then the outer decoder uses these decoded blocks to form a superblock which is decoded by its rules. This process of making preliminary decisions in the inner decoder is responsible for the loss in information which entails the penalty associated with concatenation. The inner decoder destroys information in making its preliminary decisions. It does not pass on the whole received column to the outer decoder, but rather its estimate of what this column should have been. This is the crux of the system, being responsible at once for both the degradation in achievable performance and the simplication which enables its practical realization.

We might give a simple analogy to the concatenated decoder from the familiar business firm hierarchy. Ideally the president of a company should know all the facts and figures relevant to a particular decision he must make. In any sizeable firm this is impossible. On the lowest organizational level the employees pass on their information to their supervisors. The supervisors collect this information and make preliminary decisions concerning what should be passed on to the next management level. Finally the president is presented with a simple set of alternatives and associated facts. Like concatenated coding, the business hierarchy works in principle. There is, however, clearly a loss of information as decisions are made throughout the management hierarchy.

The fact that concatenated codes of the Forney variety are conceptually useful depends strongly on the existence of a powerful, algebraic code for the outer, superblock, encoding. Fortunately a class of codes, known as Reed-Solomon codes are constructed algebraically just like the BCH codes. The code words may be represented by polynomials, but the coefficients of the polynomial, like the characters in the code, are nonbinary. Because of their albegraic structure, these codes can be coded and decoded by algorithmic procedures.

Unlike a BCH code, the error correcting ability of a Reed-Solomon code increases in proportion to its length. (This is the property necessary to obtain ideal asymptotic behavior.) The reason the Reed-Solomon codes are useless (asymptotically speaking) by themselves, i.e., without a concatenated inner code, is that the while the code length increases, so does the character size. In other words if we double the length of a Reed-Solomon code, we double the error correcting ability in characters. However as the length is doubled the size of the building block characters must increase by one bit. This is illustrated in an example sequence of longer codes shown in Figure 12. This increase in character size means that the overall ratio of *bit*-errors corrected to total *bits* does not stay constant with increasing code length. For this reason the Reed-Solomon codes are not asymptotically useful for channels where the errors occur randomly in the constituent bits.

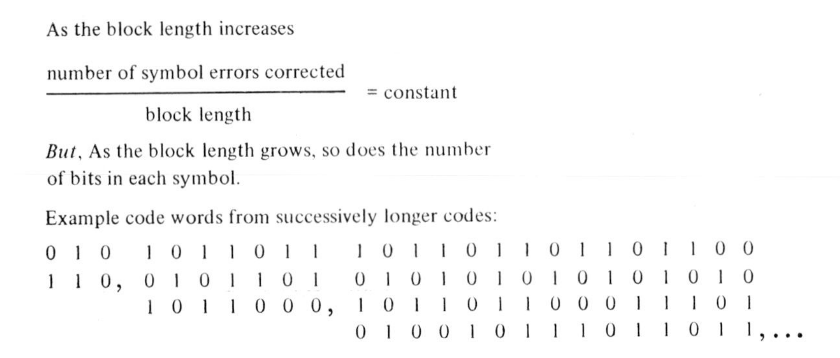

Figure 12 Asymptotic Error-Correcting Ability of Reed- Solomon Codes

The reason the concatenation scheme is effective is that the inner code can ideally convert a channel where errors occur randomly in *bits* to a channel where errors occur randomly in *characters*. For this latter channel the Reed-Solomon codes are then useful. This phenomenon is illustrated in Figure 13. We depict an example superblock in the form of a matrix of bits. The columns in this matrix are the words of the inner coder. Because errors in the communication channel occur randomly in bits, I have sprinkled errors (marked with X's) uniformly in this matrix. Notice that a Reed-Solomon decoder would be faced with 5 character errors to correct in a superblock of 11 characters were it to decode by itself without the aid of an inner decoder. This would undoubtedly exceed its error-correcting ability and the superblock would be wrongly decoded.

Now suppose an inner code is used capable of correcting any single error in one of its words. Then, as shown in Figure 13, all but one of the character errors are corrected by the inner code. It fails on the 8th word (column) where there happen to be two errors. In decoding this word wrongly it would generally introduce additional errors in the offending column. These additional errors do not matter to the outer decoder, since they all occur in the same character, which is now wrong anyway. This would be analogous to the word-by-word decoding of the "loaf of tread" sentence. Taken by itself "tread" might well be understood to mean "treat," but when "treat" is inserted in the sentence the redundancy present in the "loaf – treat" combination permits correct understanding of the information.

Initial Code Block

0	1	0	1	1	1	0	0	1	0	0
0	1	X	0	1	0	1	X	0	1	X
1	0	0	1	1	X	1	1	1	1	0
0	0	1	0	1	1	0	X	0	1	1
X	1	1	0	1	0	0	1	1	1	1

After Inner Code Correction

0	1	0	1	1	1	0	0	1	0	0
0	1	0	0	1	0	1	X	0	1	0
1	0	0	1	1	1	1	X	1	1	0
0	0	1	0	1	1	0	X	0	1	1
1	1	1	0	1	0	0	1	1	1	1

Figure 13 Error Correction in a Concatenated Code

Finally in Figure 13 the outer code can successfully decode the single character error remaining after inner decoding. Observe that the action of the inner coder has been to *redistribute* the errors. Initially the errors are random in bits; afterwards they are random in characters.

There is one trouble with this beautiful marriage of the two codes. The inner code must itself be ideal in the Shannon sense.

Otherwise it will break down as its length is increased and begin to pass on far too many character errors to the outer decoder. Thus it does not appear that the problem is solved at all. We started looking for an ideal code and ended by finding an ideal coding scheme which depends upon having an ideal code!

But we do know of one ideal procedure – the random code. The reason the random code could not be used was because of the exponential growth of the dictionary size. However, in the concatenation scheme the size of the inner code grows only logarithmically with the overall length of the composite code. Therefore, even though the dictionary size of the inner code grows exponentially with the inner code length, it grows only algebraically with the length of the overall code. This is illustrated with the sequence of codes shown in Figure 14.

Complexity of docoding inner (vertical) code grows exponentially with N

But, the number of bits needed to represent the symbols only grows logarithmically with the outer block length n

Complexity of decoding outer code grows algebraically with n

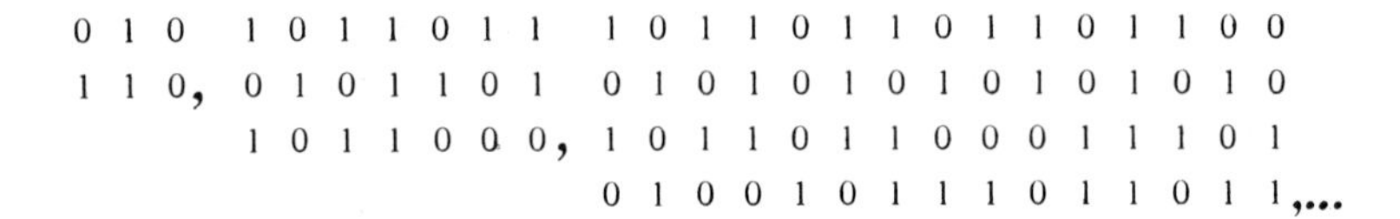

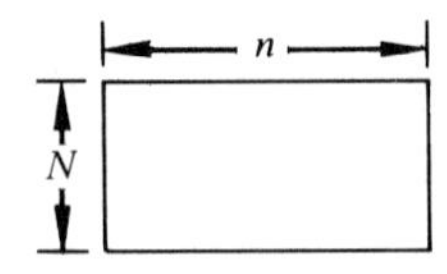

Figure 14 Growth of Complexity With Block Length in Concatenated Code

Finally we have completed what we set out to do. We have defined a coding system which can achieve arbitrarily small probability of error at any rate up to channel capacity, and for which the complexity of implementation does not grow exponentially as we approach this ideal. It may be argued that the particular scheme outlined depends upon certain "tricks" and is not therefore fundamental. Is it coincidental that other schemes which have been found for achieving ideal performance also employ concatenation? Is concatenation, or some form of hierarchical structure necessary in the construction of efficient codes? These questions cannot be answered at this time. The answers which have thus far been found are intriguing, but they whet the appetite for a more fundamental theory of mathematical structure in the attainment of Shannon's ideals.

REFERENCES

Ash, R. 1965. *Information Theory*. New York: John Wiley & Sons.

Berlekamp, E. R. 1968. *Algebraic Coding Theory.* New York: McGraw-Hill.

Bose, R. C., and Ray-Chaudhuri, D. K. 1960. "On a Class of Error-Correcting Binary Group Codes." *IEEE Transactions on Information and Control* IC-3:68-79.

Elias, P. 1954. "Error-Free Coding." *IRE Transactions on Information Theory* IT-4:29-37.

Falconer, D. D. 1969. "A Hybrid Coding Scheme for Discrete Memoryless Channels." *Bell System Technical Journal*, Vol. 48.

Fano, R. M. 1961. *Transmission of Information.* New York: John Wiley & Sons.

Forney, G. D. 1966. *Concatenated Codes.* Cambridge: The Massachusetts Institute of Technology Press.

Gallager, R. G. 1968. *Information Theory and Reliable Communication.* New York: John Wiley & Sons.

Jelinek, F. 1968. *Probabilistic Information Theory.* New York: McGraw-Hill.

Lucky, R. W., Salz, J. and Weldon, E. J., Jr. 1968. *Principles of Data Communication.* New York: McGraw-Hill.

Peterson, W. W. 1961. *Error-Correcting Codes.* New York: John Wiley & Sons.

Pinsker, M. G. 1965. "Complexity of the Decoding Process." *Problems of Information Transmission* (in Russian) 1:113-16.

Reed, I. S., and Solomon, G. 1960. "Polynomial Codes over Certain Finite Fields." *Journal SIAM* 8:300-304.

Shannon, C. E. 1948. "A Mathematical Theory of Communication." *Bell System Technical Journal* 27:379-423; 623-56.

——. 1959. "Probability of Error for Optimal Codes in a Gaussian Channel." *Bell System Technical Journal* 38:611-56.

Wozencraft, J. M., and Reiffen, B. 1961. *Sequential Decoding.* New York: John Wiley & Sons.

Ziv, J. 1966. "Further Results on the Asymptotic Complexity of an Iterative Coding Scheme." *IEEE Transactions on Information Theory* IT-13:168-71.

■

Comment on Patterns in Social Events

Magoroh Maruyama*

When one moves from the realm of physical events to the realm of social events, additional complications emerge not only from the patterns of events themselves; but also in the relationship between the pattern of events and the pattern in the perception of the observer. The first is obvious and well-recognized. The second is relatively little recognized even among social scientists. Since it contains a level structure, let me elaborate on this point.

The discrepancy between the event pattern and the perceived pattern may occur on several different levels (which I do not intend to rank-order): (a) epistemological level; (b) cognitive level; (c) focal level; (d) modal level; (e) validity level; (f) rapport level.

Epistemological discrepancies occur when the structure of the observer's model does not correspond to the structure of the event. For example, the observer may assume that a leadership structure exists in every group and proceed to analyze a basically non-hierarchical (bossless) group. He may invent criteria or measures of leaderness, apply them to the individuals in the group, and come up with an "answer" as to who is the leader. But what he is measuring may not be what he thinks can be inferred from the measurement. As another example, there was a prison inmate who was adept at listening to others' problems (like a psychiatrist). Guards observed that many other prisoners came to talk to him. Guards decided that he must be the gang leader and subsequently held him responsible as the leader of a riot with which he had nothing to do. Another example: two individuals acted as a team in shoplifting. One of them was a fast runner, and the other had a car, that is, they formed a division of labor. The runner shoplifted while the other cruised around to pick him up. On the record, the shoplifter was listed as the principal offender while the driver was listed as an accessory. Another example: Many civic administrators considered the 1967 riot in Detroit as having been caused by a small number of "leaders" or "agitators," whereas no such structure existed among the rioters and looters. Another example: In traditional Navaho society a young man may seek advice from an experienced old man without having to obey the advice. The old man's position is that of a counselor without coercive power, but unsophisticated outsiders would tend to see such a man as a leader.

Cognitive discrepancies occur when some aspects of the data do not make sense to the observer or contradict the part of the data which the

**Department of Sociology, California State College, Hayward, California 94542.*

observer has already accepted as valid. In this case the contradictory data are swept aside as insignificant or accidental irregularities.

Focal discrepancies are a matter of focus of attention. This corresponds to Dr. Wilson's remark on the choice of parameters to achieve closure (this volume). For example, one may observe the relationship between postal clerks and the postmaster from the point of view of their activities and discover a pattern. But the same relationship can be studied from the point of view of each clerk's feelings toward other clerks or toward the postmaster. The result may produce a different pattern.

Modal discrepancies occur when the mode of interpretation produces a plausible result but does not correspond to the mode of the event. For example, in restaurants the spindle which is installed between the cook and the waitresses and used to hang the customers' orders, may be interpreted in several ways: (i) the direct contact between the cook and waitresses may produce a father-daughter complex, which is avoided by the spindle; (ii) the direct contact between the cook and the waitresses could produce a reversal of the organizational hierarchy, in which the lower-paid waitresses give orders to the higher-paid cook. This reversal is avoided by the spindle. Regardless of these interpretations, the actual primary purpose of the spindle may be to spatially place the orders in a time sequence. The sequence can be inspected from both sides by simply rotating the spindle. Moreoever, as old orders are filled and new orders come in, the spindle can be rotated, avoiding the necessity to move up the orders one by one. Another example is the interpretation of the use of violence among ghetto residents. Some sociologists often speculate that violence is a means to achieve acceptance among peers. My findings indicate the opposite. Children in a ghetto learn to fight to defend their lunch bag, money or groceries *against* their peers. Parents encourage their children to fight back, not to be accepted by the neighbors but in order to defend their groceries against their neighbors. The child who does not fight is not rejected, but welcomed and exploited.

Validity discrepancies occur when the researcher does not get his feet wet in the reality of events. For example, some researchers base their work on official crime records which are often superficial, distorted or inconsistently classified.

Rapport discrepancies which interest me most, occur when the information giver produces phony information to satisfy the data collector. This can happen in several ways: (i) when the information may produce effects harmful to the information giver; (ii) when the research is perceived as useless or irrelevant to the information giver's side, while being beneficial to the side (considered to be an enemy) such as a collector of tribal folklore from a museum in a white society; (iii) when the research is perceived as a personal tool for the researcher, such as gaining a professional reputation, income, academic degree, or proving his hypothesis; (iv) when the researcher is perceived as sincere but naive and is not fully aware of or sensitive to the danger, harassment, abuse and exploitation which may result from the information itself or from the act of giving information, and consequently the researcher may pass information unintentionally onto the wrong side. Prison inmates, for example, have different sets of information to give to different persons: one set for newspaper reporters, one set for prison psychologists, one set for social science researchers, etc. Their survival depends on their ability to handle information, and they are expert at this art. My current thinking is that rapport is obtained, not by "techniques" of anxiety reduction, but by making the purpose of the research converge to the purpose of the researched.

I have been conducting projects in which in-culture persons conduct their researches not as mere data collectors but as epistemological conceptualizers, focus selectors, hypothesis makers, methodology designers and data interpreters.

■

Levels and Integrated Entities

Bill Wells*

The importance of the notions of integration, complementarity and integrity applied to social organization leads us to search for basic ideas underlying integrative forces and processes observed in nature. We have identified five principles in the nature of integration which suggest criteria for evaluating proposed modifications to our social order. These are:

1) *The structure of reality is hierarchical.* All physical entities can be classified according to their level of structural complexity, beginning with the elemental particles as the first or simplest level and progressing to atoms as the second level of complexity, then on to molecules, cells, multi-cellular life forms, several levels of social entities, and so on up to cosmic entities and aggregates.

2) *The number of entities from one level that can be integrated to form one entity of the next higher level is limited.* The maximum number of electrons that will "stick together" with the various other components in one atom is something less than 100. With the exception of crystals, molecules are generally of submicroscopic size and are composed of limited numbers of atoms. The living cells of which all living things are composed are microscopic in size. Even though they are composed of complex hierarchical structures of molecules they are too small to be visible without special equipment. This limiting principle applies on all levels and has long been recognized by organizers of enduring social structures. Principles of military organization have pretty well identified optimum values, so that no squad will have too many men, no platoom too many squads, no company too many platoons, etc.

3) *There is a complementary relationship between and among the components of an integrated entity.* Components of an integrated entity are not identical; they must differ such that each provides something the other needs or share something both need. Two atoms of fluorine share an electron and become a molecule of fluorine. Positive sodium and negative chlorine neutralize their differences, and become salt. It is electrical differences which bind atoms together to form molecules. In machines the differences are usually structural while in social groups the complementary difference is often functional. It is the differences between components that provide the binding forces of integration which hold the entity together, whether it be an atom, family, work crew, or community.

4) *All components of an integrated entity are integral and necessary to its completeness.* Although the word "integrate" is often used to replace

**105 NW Ninth Street, Oklahoma City, Oklahoma 73102.*

"desegregate" so as to imply a more positive action toward social "putting back together," integration definitely does not mean "to mix." The significant difference is that mixing produces "stuff," but integration produces "things." We mix sand, water, and cement to produce concrete. We can shape the concrete into blocks or other forms which we may call entities; but they are not "entities" in the same sense that a watch, an atom, or a military company is an "entity." We can throw away part of the concrete without changing the functional character of the remainder, but if any part of the watch is missing, its function is altered if not entirely disrupted.

5) *Entities which are not components of entities of a higher order are the dynamic elements of the universe.* These are commonly called "free;" free electrons, free atoms, free molecules (a virus), or free cells (those causing disease). A better name for this type of entity might be "primary" because it constitutes the highest complexity of its own particular level. In other words, since it is not a part of something bigger than itself, it is available for integration. Communism recognizes the necessity to disintegrate a non-communist society in order to make its components (people) available for integration into the communist pattern.

We have made only a few elemental observations about the nature of integration – not sufficient compared with what yet remains to be learned. But these observations suggest certain criteria for evaluating some of the things we are attempting to do on the social level.

■

Structural Hierarchy: The Problem of the One and the Many

Ronald G. Jones*

From approximately 3,000 B.C. in the East when Emperor Fu Hsi is believed to have discovered the idea of Yin and Yan as they are embodied in T'ai Chi, and since about 400 B.C. in the West when Heraclitus and Plato struggled with the problem of opposites, scholars have tried to resolve the mystery of how many parts come to be a unified whole. This adds up to a 5,000 year old problem. Throughout the history of ideas the problem of the One and the Many, of part-whole relations, of order and structure has appeared over and over again. Names we are all familiar with identify this as the *pons asinorum.* Nicholas of Cusa (15th Century), Giambattista Vico (17th Century), and more recently Cassirer, Whitehead, von Bertalanffy, Koestler, Sorokin, and Polanyi have made this problem *the* problem. And this is to list only a few. From among us at this symposium, Lancelot Law Whyte has written that the principle of the union of contrasts relates directly to the most urgent needs of man. The problem of the integration of differentiated parts, of harmony in diversity, is not merely a problem for idle, remote, and academic speculation.

This is what I take to be the concern of this symposium – the interdisciplinary search for "the idea of structural hierarchy" – a problem with at least a 5,000 year history. But this conference adds a dimension not always present in earlier attacks on this problem. Indeed, this dimension could not have been present, for it requires the highly specialized findings of modern science. Perhaps it will be the contributions from the fields of modern physics and biology that will help us in the personal and social realms.

To make certain that this connection is clearly before us, let me elaborate a bit. It seems to me that the *sine qua non* of man's knowledge, happiness, and existence is to be found in the idea of the reconciliation of differences. It matters little whether we talk about mental health and personality structure or whether we talk in the context of society. It matters little what the size of the society is. It makes little difference whether the society is a marriage, a small group, a large industrial organization, a community, a nation, or many nations, the basic issue is that of the reconciliation of the individual with the group, the organization, the integration of parts into a unified whole. These issues are all matters of totality, wholeness, completeness, unity, order, structure.

**Department of Educational Foundations, Faculty of Education, University of British Columbia, Vancouver 8, British Columbia, Canada.*

If, as Dr. Smith claims in his work on crystal structure, the nature of the entity derives from the system in which it is embedded – then the context reflects back on the nature of the part. This means that in addition to the autonomy of parts there are also reciprocal or internal relations. Of course, philosophers have argued this for a long time, but it is something else again to have such beautiful empirical evidence, and it adds to the solid basis for understanding the harmonious subordination, superordination, and coordination of parts within a whole.

In a recent bulletin of the Association of Universities and Colleges of Canada, *University Affairs*, the lead article states that "Quebec completes structure for universal post-secondary education." The article goes on to say that the system of 1969 bears little resemblance to the structure of 1965. For me the important question revolves around the terms "system" and "structure." What kind of hierarchical structure? There are alternatives, and it makes a difference as to the model used. Personality structure, social structure, political order, international order, and the educational system that goes with them, require understanding of the hierarchical arrangement of parts. But in the final analysis, we have never understood this problem of structure beyond a few well-worn ideas such as "parts are interdependent." This does not take us very far.

I am reminded of statements by Scott Buchanan that typify for me the challenge of this symposium. Like the turning point in tragedy, there is a reversal or conversion where everything said moves to a different level and is transfigured. "Reversal or conversion in sublime discourse happens when speech or writing reveals second intentions," and this awakens a sense of glory. It seems to me that "the idea of structural hierarchy" can help in awakening a sense of glory and hope, since solutions to the problems of fragmentation on the many levels of today's world rest on just such knowledge. The One and the Many – once again.

■

Bibliography

Forms of Hierarchy: A Selected Bibliography

Donna Wilson*

Herein we deal with the subject of hierarchy in three broad categories: hierarchy as concept, hierarchy in nature, and hierarchy in artifact. Many have remarked on the ubiquity of hierarchy – the observation that both natural and artificial phenomena are structured in levels and sets of ordered levels. It is not a new idea nor unique to the references cited here. What is novel in this provisional selection is the juxtaposition of the various disciplines and their specific content under this single theme. This effort basically supports the theme of this symposium, "Hierarchical Structures in Nature and Artifact," although we have not limited citations to this subject alone. Our purpose is to suggest a direction for scanning literature from many diverse sources and to outline possible areas for further search. Since we can make no claim for completeness nor have as yet a valid measure of relevancy for any one entry, additions and correction to this compilation would be greatly appreciated.

HIERARCHY AS CONCEPT

One key paper in this category is Simon's "The Architecture of Complexity" (Simon 1962). His central theme is that "complexity frequently takes the form of hierarchy," and that "hierarchic systems have some common properties that are independent of their specific content." Simon poses a definition of hierarchical systems and explores reasons for hierarchical organization. His definition of hierarchical system is a system "composed of interrelated subsystems, each of the subsystems being in turn hierarchic in structure until we reach some lowest level of elementary subsystem." In discussing the varieties of hierarchy from crystal structure to social systems, Simon defines the *span* of a system as "the number of subsystems into which it is partitioned." *Flat* hierarchies such as a volume of molecular gas or a diamond are those where the ratio of the number of levels to the span is small (e.g., $1{:}10^{10}$) in contrast to steep hierarchies, where the ratio of the number of levels to span is large (e.g., 1:10).

Simon's paper includes the wonderful image of the two watchmakers, Hora and Tempus, who demonstrate the advantages of modularization. Hora builds watches in modules while Tempus assembles watches element

**Douglas Advanced Research Laboratories, McDonnell Douglas Corporation, Huntington Beach, California, 92647.*

by element. Hora prospers while poor Tempus eventually goes out of business. Why? The reason is found in the fact that although both are interrupted by phone calls and customers, Hora need not resume each time from scratch. In a study that discusses modern man's predicament in studying himself, Koestler expands this tale to the point of suggesting that life itself is possible only because of its hierarchic organization (Koestler 1967). The advantage of modularization thus induces Simon and subsequently Koestler to argue that complex systems evolve far more quickly when hierarchically organized.

Another property of hierarchic systems is that they are "nearly decomposable," that is, interactions among subsystems are relatively weak compared with interactions within subsystems. This facet not only greatly simplifies their behavior, but it greatly simplifies the description of complexity (Simon 1962). Weaver outlined the study of complexity in 1948 when he called for greater effort to study problems of a middle realm between problems of few variables (simplicity) and those of many variables (disorganized complexity) by utilizing the new holistic methods of systems analysis and operations research (Weaver 1948). In addition to Simon's similar bid for a more substantive approach to the study of complexity, Wilson points out that bigness as well as complexity is treated hierarchically in the natural order. "Direct confrontation of the large and the small is avoided in nature. A hierarchical linkage is always interposed. Bigness is avoided in the sense that the ratio between the size of the structure and the modules out of which it is built is functionally bounded." (Wilson 1967).

Before going further into this category, hierarchy as concept, we need to review the context of discussion and find what vocabulary already exists in the literature. Koestler addresses the question of what to call entities that belong to hierarchical systems. They have two aspects, ". . . the functional units on every level of the hierarchy are double-faced as it were: they act as wholes when facing downwards, as parts when facing upwards." (Koestler 1967). He elects to designate these "Janus-faced" entities by the term *holon* (from the Greek *holos* = whole plus the suffix *on* as in proton or neutron suggesting a particle or part). We note that Gerard uses the term *org* to designate the same concept (Gerard 1957).

Whyte traces the idea of hierarchy historically and finds that the concept of "a sequence of higher and lower levels" has been important in western thought since Plato (Whyte in press). By 1900 the term *hierarchy* was used both for taxonomy and for classifying forms of energy; however, the importance of "hierarchical relationships for biological theory" has had a growing recognition only since 1910. Whyte argued in 1949 that "a great hierarchy of relations of dominance guides the differentiation of the developing embryo. . . ." Thus, the concept of level is not only an old idea, but it is necessary for describing much that is observed in the universe, and even though "the obvious is hard to analyze," we must refine and sharpen what we mean by *level,* a definition basic to the concept of hierarchy.

Bunge addresses the subject of hierarchy by enumerating the use of the term *level* in contemporary science and ontology (Bunge 1959, 1960, 1963). He examines nine meanings of the notion of level and raises questions in connection with each. He considers category number nine to be an adequate definition of level, that is, "grades of being ordered, not in arbitrary ways but in one or more evolutionary series," and suggests that it is this meaning that is intended in the idea of *level of organization* (Bunge 1963). Bunge's category four, called emergent whole, is the concept employed by biologists and psychologists to convey the notion of lower order wholes becoming the building blocks of higher order wholes. His descriptive scheme of these nine categories is shown in Figure 1, and he limits the use of the notion of level to include both the idea of emergence in time without restricting the direction (i.e., both lower to higher and higher to lower) and the fact that level structure need not be restricted to linear gradation – it can be parallel, branched, etc.

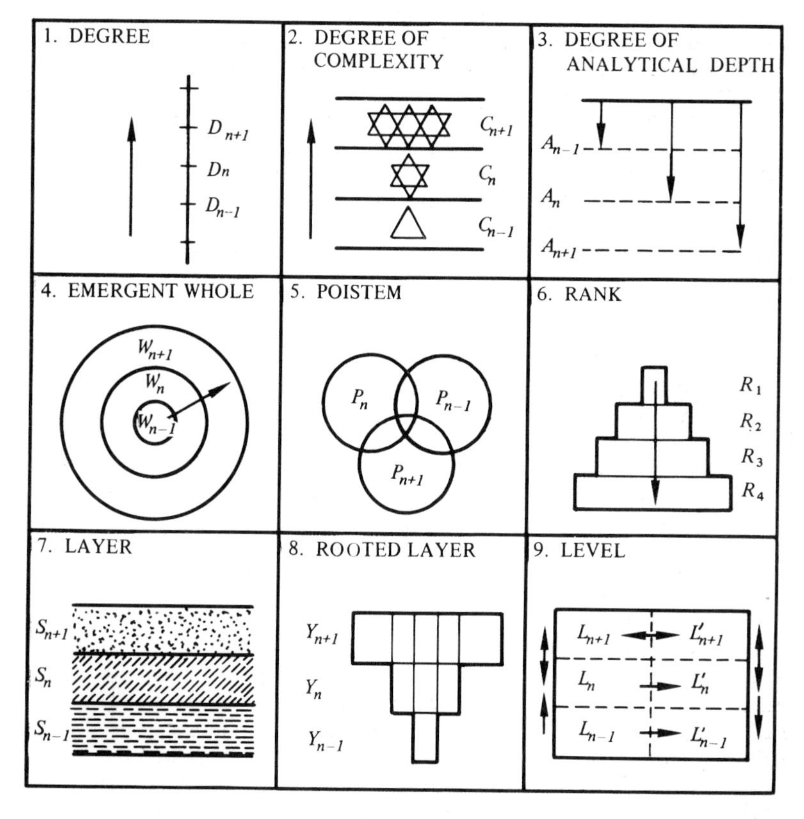

Figure 1. Uses of the Term Level
From "Levels: A Semantic Preliminary." (Bunge 1960a) by permission of the author.

Continuing the discussion of the level structure of reality, Marjorie Grene (1967) examines the question of whether "... a one-level ontology [is] adequate to account for the major areas of human experience, ..." and, if not, how is it possible to formulate a many-level ontology? Although her discussion mainly concerns the philosophical foundations of biology, i.e., the problem of reducing all biological explanation to the level of physics and chemistry, she takes the subject into areas that question whether even physics and chemistry are molecular sciences. The importance of hierarchical notions to the question of reductionism in biology is central in Woodger's (1952) *Biology and Language* whose subtitle is "An Introduction to the Methodology of the Biological Sciences including Medicine." Woodger's formulation of the methods of biological inquiry in the language of set theory and symbolic logic will perhaps deter many from a detailed reading, but it is important to realize that the current status of our understanding of hierarchical structure in living phenomena lacks an adequate mathematics. "One of the future tasks for biological methodology is the discovery of the kind of mathematics that is required for biology. Considerable use has already been made of some existing branches of mathematics, but these branches have been developed, to a very large extent, to meet the special demands of physics." (Woodger 1952). A more understandable exposition of the logical notions useful in describing level-structure phenomena is found in a discussion of the taxonomic Linnaean hierarchy, "a system of nested classes whose members are individual organisms" (Buck and Hull 1966).

So far we have not found any substantive discussion in the literature concerning the forms of hierarchy other than Hawkin's remarks on the idea of a *scala naturae*, a "ladder" of nature in contrast to the Darwinian evolutionary "tree" (Hawkins 1964). The hint that hierarchy is not limited to the form of pyramids or trees is found in Alexander's paper, "A City is Not a Tree," where he argues for a semi-lattice arrangement in the design of cities, although he does not expand on the concept of hierarchy (Alexander 1965). Smith (1964) differentiates cellular aggregate from branched structure in crystals. Bunge (1963, 1967) restricts hierarchical form to sequences of terms ordered by a one-sided (asymmetric dependence) relation. Maruyama discusses the form used to organize information and argues that hierarchical arrangements are limited. Hierarchic organizations of information derive from conceptualization modes that he calls *classificational* in contrast to *relational* or *relevantial* (Maruyama 1965). Because many of the specific references cited in this bibliography do not emphasize hierarchy *per se* (some do not actually recognize the hierarchical aspect of their subject even when it is implied), it is best to go directly to the examples found in the literature. Our organization into the particular subheadings within each of the next two sections demonstrates the variety of hierarchical form.

In the section, *Hierarchy in Nature,* we consider works whose subjects are 1) physical, that is, fundamental particles, molecules, crystals and cosmic aggregates; 2) biological, from virus structures to the human nervous system; and 3) geometrical, aggregates resulting from close-packed polyhedra. Under *Hierarchy in Artifact,* we consider 1) software, such as codes, languages, programs and search strategies; 2) hardware, including computers, transportation systems and cities; 3) organizations, such as files, data processing, management schemes and social systems; 4) cognition, which deals with levels of knowing, memory and pattern recognition; and 5) epistemology, which cites both classical and modern attempts to classify knowledge and disciplines.

HIERARCHY IN NATURE

Purcell (1963) discusses problems from physics that are characterized by the need "to understand the behavior of the aggregate in terms of the elementary laws governing its individual parts." Certain transitions between order-disorder states of aggregates cannot be explained in terms of the known properties of their component parts and interactions. Two examples are discussed in detail: (*i*) the Ising model of magnets, and (*ii*) the Adler/Wainwright computations utilizing a "billiard-ball" molecular gas model. Both examples are worth looking up for the insight they afford into the "stubbornness" (Purcell's term) of these order-disorder transitions. Roosen-Runge offers an approach to parts and wholes in a mathematical discussion of the "logical relationships between the specification of a part and the description of the whole to which it belongs." (Roosen-Runge 1966). A more complete review of his paper than is possible here is necessary to discuss his formulation; however, its relevance to our subject is noted.

Weiss in a beautifully illustrated paper (Weiss 1967) discusses many of these same problems as "the progression from elements to groups." One central thesis is that as we go down the ladder between telescopic and microscopic vision; we gain precision but lose perspective. What is lost in decomposing wholes into parts is "plainly the interrelations that had existed among the parts while they were still united." In order to reconstruct wholes from these decomposed fragments, it is necessary to add a descriptive term "that specifies the lost relations." He further argues that the solution to reductionist thinking which results from isolating component parts from their context is to be found in adopting an attitude of holism. This holistic attitude must look for the formative behavior behind the appearance of "static geometrical regularities of pattern." He asks, how does the spiralness of galaxies derive from the properties of stars and the cyclonicness of cloud patterns from the properties of aerosols? In conclusion, he argues that "we must realize that individual freedom in the small is compatible with the existence of collective order in the gross. . .self-patterning of groups occurs among molecules and men alike." (Weiss 1967).

From the discipline of crystallography, Smith differentiates two basic forms: *cellular* structure that is illustrated in crystalline aggregates and foam (soap bubbles), and *branched* structure (tree-like) that is illustrated in electric discharge, corrosion and crystal growth (Smith 1964). He points out that cellular systems are "constrained toward a minimum area of interface" while branched structures "derive from the growth of isolated individuals – this occurs whenever a protuberance has an advantage over adjacent areas in getting more matter, heat, light, or other requisites for growth." We note in passing that what Smith calls "branched structure" is referred to as "anastomotic" networks in mathematical modeling of neuron organization (Hawkins 1967). Smith refers to hierarchy when he says "repeated or extended irregularities in the arrangement of atoms become the basis of major structural features on a large scale, eventually bridging the gap between the atom and things perceptible to human senses." (Smith 1964). In asking why science cannot develop a new approach to encompass the extremes of atomistic physical chemistry and averaging thermodynamics, he reminds us that, "It is neither possible nor necessary to study all structures that might have existed, but there is need for studying more than a statistically averaged structure." (Smith 1968).

Still within the realm of physical hierarchies in nature, we turn to the subject of hierarchy in the cosmos. A hierarchy of satellite systems was first proposed by Lambert in 1750 (Wilson 1965). A solution to the Cheseaux-Olbers' paradox (which states that if stars are more or less of the same intrinsic brightness and distributed more or less uniformly, the night sky should be as bright as the sun) is found by introducing a hierarchically structured universe (Charlier 1922). Although a hierarchical distribution can account for the observed night-sky brightness, it is not widely used in cosmological models (Harrison 1965, Wilson 1965).

In another astronomical discussion, Weizsäcker (1951) poses the difficulty of understanding evolutionary processes required by cosmological theories. He formulates the problem as one of parts and wholes: "The evolution of a single object can be understood only in terms of its temporal and spatial boundary and external forces acting on it. These conditions, however, are defined by the evolution of the large system of which the object forms a part." Although he does not expand this dilemma in his technical discussion that basically is a turbulence theory for the origin of galaxies, Weizsäcker utilizes both the concepts of a "hierarchy of eddies" and a "hierarchy of clouds." Another astronomer has pointed out the hierarchic organization of nature many times and his familiar chart that classifies material systems is reproduced in Table I (Shapley 1958). A concern to find "man's place" is reflected in a scale that locates the microcosmos to the left of man (negative numbers) and the macrocosmos to the right (positive numbers).

In biology we find numerous discussions concerned with the emergence of life at some level within a sequence of levels. Palade calls this arrangement of living systems that are composed of a relatively few common chemical elements "a hierarchy of structural patterns." (Palade 1963). His review of cellular formation and structure points out that despite the fact that there exist "far-reaching distinctions" among cells, there is "no structural unity at the cellular level." Bertalanffy (1952) also discusses the inadequacies of the "cell theory" but argues that "the cell of a unicellular organism is homologized *only* with the multicellular organism as a whole, not with its individual cells" (italics mine). Bertalanffy further discusses the hierarchical pattern of biological organization and amplifies the principal of hierarchical order defined and analyzed by Woodger (1937).

Kellenberger (1966) summarizes what is known about the structure of viruses and how their shape is genetically controlled. Viruses can be characterized by shape (a well-determined shell of protein) and hereditary information (a core of nucleic acid). The protein shell, which is an assembly of subunits, has the shape of an icosahedron in some viruses, i.e., those with shells of sixty subunits or less. The shape of more complicated viruses has not yet been determined, and the mechanism of shape-making is not understood. Caspar and Klug (1962, 1963), working on why icosahedral shape, find reasons having to do with bonding properties and energy that go beyond geometrical regularity (Kellenberger 1966). However, Kellenberger also concludes in a similar vein to Palade, "Knowledge of the genetic control of shape in protein structure will not be enough to explain the origin of shape in higher organisms. In multicellular systems, cells differentiate into specialized groups. The shape of such an organism depends on the differential growth of specialized cells, and that growth is regulated in part by the interaction of cells." The problem of parts and wholes is common to biology as well as physics.

Other discussions included here on the biological problems of parts and wholes are Gerard (1958), an edited proceedings of a symposium; Redfield (1942), an introduction to a symposium that discusses other relevant papers; and Prosser (1965), a recent summary including a critique of the concept of information theory applied to problems on evolution. Each is concerned with levels in biological organization and includes extensive references to the literature. The concept of integrative levels in biology is examined by Novikoff, who objects more to a stretched analogy between

Table I. A Classification of Materials

–5 . . . . .

–4 Corpuscles (Fundamental Particles)
- α. . . . . . .
- β. Radiation quanta
- γ. Electrons
- δ. Protons
- ϵ. Neutrons
- ζ. Positrons
- η. Mesons, 1 to x
- θ. Neutrinos
- ι. Antineutrinos?
- κ. Antiprotons
- λ. . . . . . .

–3 Atoms
0 to 101+

–2 Molecules
1 to n

–1 Molecular Systems
- I. Crystals
- II. Colloids

0 Colloidal and Crystallic Aggregates
- α. Inorganic (minerals, meteorites, etc.)
- β. Organic (organisms, colonies, etc.)

+1 Meteoritic Associations
1. Meteor Streams
2. Comets
3. Coherent Nebulosities

+2 Satellitic Systems
- I. Earth–Moon Type
- II. Jovian Type
- III. Saturnian Type

+3 Stars and Star Families
- α. Stars with Secondaries
  - I. With Coronae, Meteors, and Comets
  - II. With Nebulous Envelopes
  - III. With Planets and Satellites
- β. Stars with Equals
  - I. Close Pairs (or Multiples)
    - a. Eclipsing
    - b. Spectroscopic
  - II. Wide Pairs (or Multiples)
    - (α) Gravitational
    - [(β) Optical]
  - III. Motion Affiliates

+4 Stellar Clusters
- α. Open
  - [a. Field Irregularities]
  - b. Associations
  - c. Loose Groups
  - d. Compact Groups
  - e. Dense Groups
- β. Globular
  - I. Most Concentrated
  - II. . . . . . .
  - . . . .
  - XII. Least Concentrated

Table I. (Cont)

+5 Galaxies
- A. Bright
  - I. Irregular (I)
  - II. Spiral (S)
    - α. Abnormal (Sp)
    - β. Barred (SB)
      - (1) Open (SBc)
      - (II) Medium (SBb)
      - (III) Concentrated (SBa)
    - γ. Regular (S)
      - (I) Arms Very Wide (Sd)
      - (II) Arms Wide (Sc)
      - (III) Arms Close (Sb)
      - (IV) Arms Very Close (Sa)
  - III. Spheroidal (E)
    - a. Most Elongated (E7)
    - b. Less Elongated (E6)
    - . . . .
    - g. Least Elongated (E1)
    - h. Circular Outline (E0)
- B. Faint (Bruce Classification)
  Concentration and Shape
  a1 a2 a3 . . . a10
  b1 b2 b3 . . . b10
  . . . .
  f1 f2 f3 . . . f10

+6 Galaxy Aggregations
1. Doubles
2. Groups
3. Clusters
4. Clouds

[5. Field Irregularities]

+7 The Metagalaxy
- α. Organized Sidereal Bodies and Systems
  1. Meteors
  2. Satellites
  3. Planets
  4. Stars
  5. Clusters
  6. Galaxies
- β. The Cosmoplasma or Matrix
  - (α) Interstellar Particles
    1. Cosmic Dust and Meteors
    2. Diffused Nebulosity (dark)
  - (β) Interstellar Gas
    1 Corpuscles
    2. Atoms
    3. Molecules
  - (γ) Radiation
  - (δ) . . . .

+8 The Universe: Space–Time Complex

+9 . . . .

Subdivision symbols:

α, β, γ . . . . differences largely dependent on basic nature
0, 1, 2, 3 . . . . differences largely dependent on size or mass
I, II, III . . . . .differences largely dependent on structure
A, B, a, b . . . . Differences largely dependent on position of observer

The three groups in square brackets are chance associations, not graviational systems.

From: *Of Stars and Men* (Shapley 1958), by permission of the author.

society and living organisms than to whether the concept itself is adequate (Novikoff 1945). In a rebuttal, Needham citing Woodger as the pioneer of this concept makes the following statement about the concept of level: "Once we adopt the general picture of the universe as a series of levels or organization and complexity, each level having unique properties of structure and behavior, which, though depending on the properties of the constituent elements, appear only when these are combined into the higher whole, we see that there are qualitatively different laws holding good at each level." (Needham 1945). Needham's fuller development of the subject of integrative levels and organization is found in his *Time: The Refreshing River* (Needham 1943). A critical review of level organization utilized by organismic biology is given by Beckner (1968). He argues, "The world is not constituted of neatly separated strata that force certain distinctions upon anyone who observes it without preconceptions. Rather, bits of the world break into strata when a class of phenomena are approached by an investigator equipped with a set of concepts. . . . The problem of the selection of a level of analysis can arise when a set of phenomena is not understood." However, Bradley (1968) attempts to analyze biological processes as multi-level systems just because of the persistent failure to solve then at the molecular level. He applies a multi-level systems analysis to three molecular and submolecular level models of biological processes – RNA and memory, DNA replication and codon-anticodon recognition. The problem of levels in biological systems is far from settled. Indeed, even with the advent of cybernetic and information theoretic contributions to biological understanding (Prosser 1965), a state of confusion still exists and our ignorance of formative processes observed in nature (Whyte 1965) remains.

In a paper first translated into English in 1964, Khailov discusses the application of general system theory (Bertalanffy 1955) to theoretical biology. General systems theory considered in the light of the classical theory of evolution reveals three problems: one is to define "living system" in systems terms; another is to enumerate objects that can be studied in their systems aspect; and the last is to establish a hierarchy of living systems (Khailov 1964). On the last, Khailov raises the issue of the position of macrosystems on the hierarchical ladder. Differentiated macrosystems from the organism up to the ecosystem, etc., are "connected by inclusion, that is, each is included in another system and is open to the latter." (Khailov 1964). Bertalanffy's recent summary of General System Theory (1968) provides both a historical view of the foundation and development of this approach and outlines further areas of research that are central to the subject of hierarchies.

Both Mesarović (1968) and Rapoport (1966) continue the discussion of the systems approach and its relation to biological systems. Mesarović summarizes two principles derived from his research on hierarchical or multi-level systems. The "principle of overflow of interaction" derives from considerations of the problem of how the information flow between two levels leads to harmonious functioning of the system. It states that coordination is achieved by providing the second (higher) level with two types of information: the request for change in interactions and the dependence of the goal parameters upon changes in interactions. This is less information than is needed for complete control from the second level. The second "principle of optimal communication level" points out that excessive levels of communication as well as interrupted communication channels can disrupt a multi-level system (Mesavorić 1968). Miller also adopts the general systems view of a hierarchy of systems in his research to study information overload at several levels of living systems. He finds a similar response to information overload, that is, a breakdown under stress of a high rate of information input "whether the system in question is a neuron or a human group" (Miller 1964a, b).

In another work of major breadth, Miller defines basic concepts in living systems, reviews biological investigations of structure and process[1] and posits some 165 "cross-level" hypotheses, that is, generalizations that "appear to be true of systems at two or more levels." (Miller 1965). So far, I have not found extended discussion or review of Miller's work, a serious omission in the literature in view of the potential of such a powerful tool for synthesis.

Rosen, drawing upon the successful utilization in physics of optimality principles such as Fermat's Principle of Least Time and Maupertius' Principle of Least Action, etc., discusses how optimality principles apply in the biological world (Rosen 1967). The relevance of Rosen's thesis to hierarchically organized systems is the problem of obtaining systems descriptions at the biochemical level (Rosen 1968).

And finally, within this category, Bronson (1965) presents a neurological model to relate behavioral with neurological data. His model emphasizes the hierarchical nature of the organization of the central nervous system by postulating three "levels" within the nervous system. In a discussion of how children learn, Hawkins (1967) emphasizes the "anastomotic" structural similarity between the brain and networks in

1. A spurious dichotomy if we acknowledge that "what are called structures are slow processes of long duration, functions are quick processes of short duration." (Bertalanffy 1952).

large reliable computer systems. He draws this analogy because the redundancy provided by anastomotic networks results in the most efficient computational or classificational capability in the presence of noise. Bateson (1968) also expands on the concept of redundancy in the communication systems of humans and animals. In another work, he develops a four level structure of learning that has important consequences for changing behavior (Bateson 1960). Stewart's research on electrical fields in densely packed cellular media is concerned with understanding "details of brain mechanism and its relation to behavior." (Stewart 1963.) In addition to the 'Lillie iron-wire' nerve model used to study field phenomena, he describes results of growing electrodeposited gold dendrite trees and of observing conduction in simulated cellular media. We note the hierarchic aspect of these structures—both the aggregate form of electrical stimulation in close-packed aggregates of small pellets submerged in electrolyte and the tree-like or branched form in the dendrite trees.

Turning to geometrical hierarchies that derive from close-packing of polyhedra, Smith (1954) compares cellular aggregates found in crystals, soap bubbles, insect wings, living cells, etc. The application of this kind of investigation to virus structure has already been noted. A review of regular polytopes (Coxeter 1963) is beyond our scope; yet the hierarchical aspect of these structures is evidenced in the repetition of form in successive "shells" or layers (Fuller 1965). To date, too little refinement and formalization of these geometrical studies of natural structures exist to allow their generalization to other fields. The value of these studies for architecture and design is realized (Burt 1966); however, this takes us into the subject of hierarchy in artifact. Since we include only two references to mathematical hierarchy (Gardner 1966), which describes Cantor's hierarchies of infinities and (Sankaranarayanan 1969), which discusses a group-theoretic connection among the hierarchical levels of physics, we mention them here rather than later under artifact. Another reference that employs hierarchical notions worth mentioning (but difficult to classify) is Leake's article on the ethical aspects of experimental studies on human subjects. (Leake 1967). He sees that part of ". . . our difficulty over ethical problems results from conditions of organizational levels of living material. Through lack of knowledge we tend to confuse the factors operating at an *individual* level of biological organization with those operating at a *social* level."

HIERARCHY IN ARTIFACT

From the field of communication systems, we find examples of hierarchy that come within our category of software. Hierarchical structure is observed in results of coding methods (Huffman 1952, Forney 1966, Lucky 1967). Huffman defines a minimum redundancy code as "one constructed so that the average number of coding digits per message is minimized." The "tree-like" structure of the schematics presented in his conclusions is hierarchical although he does not point out this aspect. The scheme of concatenated codes reviewed by Forney (1966) derives from solutions to the problem of error-correcting codes (Lucky 1967). Briefly, the problem that arises in the transmission of binary data is one of assuring the overall system capability of error correction while keeping the length of the coding and decoding implementation from growing exponentially. The only solutions found so far are to place codes within codes – a method called concatenation.

Decision-making strategies found in much of the literature on operations research and systems analysis take the form of hierarchy as well as search strategies required in file organization (Becker and Hayes 1963). In a discussion on the theories of file organization, Becker and Hayes described the model of "activity" organization which "supplements methods of logical organization. . . . The aim of activity organization is to produce a hierarchical arrangement of nested "boxes" or levels of grouping, which will represent a compromise among various 'usage' distributions in such a way as to optimize the selected measure of efficiency. . . . These sets of boxes become quite analogous to the structure of a normal classification scheme, although their method of derivation is dependent on the character of usage rather than *a priori* decision." Their diagram of nested boxes to represent file organization (Figure 2) is identical in form to Woodger's *division* hierarchy of the cell.

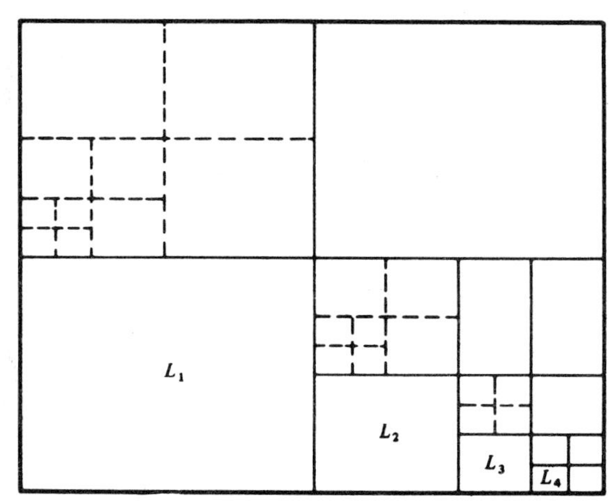

Figure 2 Hierarchical File Organization (after Becker and Hayes 1963)

The results derived from language analysis also take the form of hierarchical structure (Chomsky 1957, 1967; Koestler 1964, 1967). Chomsky's formal description of the syntactic component of a grammar is not only more clearly exposited through the use of a tree-diagram; there exists a structural homology of hierarchy between physiology and syntax. Koestler utilizes the hierarchic form of Chomsky's analysis to illustrate the point that most human skills (instinctive or learned), including active speech, cannot be adequately represented by the S-R (stiumlus/response) chain of behavorist psychology, but require the "tree-branching" process which is characteristic of all hierarchic processes.

Turning to computers, we find a statement by Von Neumann as early as 1949 on the hierarchic organization of computer storage (Von Neumann 1958, 1966). He claimed that computer memory is characterized by capacity and access rate to the storage. Because there is no known technique for building a memory with both adequate capacity and sufficiently fast access, it is necessary to organize computer storage hierarchically. His scheme for accomplishing this is to make the first memory of sufficient access rate but small capacity, to add a second memory with a much larger capacity but of slower access rate than the first, and then to add a third memory with a larger capacity but a slower access rate than the second, and so on. Evidently this principle enunciated at the onset of the computer age still holds in computer design.

In transportation systems we find examples of modular hierarchy in the concept of cargo containerization and in the structure of multi-modal systems. Implicit in Doxiadis' (1968) analysis of the structure of cities and alternative solutions to their congestion and decay are hierarchic modules that are polynucleated. Wilson establishes a homology between the maximum size of a city and the maximum size of gravitating cosmic aggregates and concludes that hierarchical modular structures provide a way to accommodate indefinite size while satisfying limitations such as density or commuting time (Wilson 1967).

On the subject of social organization, we mention only two authors, (Landau 1965 and Brams 1966, 1968, 1969) from the extensive literature that exists; however, further search in the directions of references cited in these papers would be fruitful. Specifically on hierarchical structure, Landau derives a hierarchy index to measure "nearly hierarchical" structure in societies. He mathematically treats three models of social arrangements: the *tournament model,* in which *n* members come together and engage in contests, the result of each of which is independent of any other contest and fixes the direction of dominance for the pair involved; the *Markov chain model,* which is a society of *n* members among whom dominance relations are established in some unspecified manner, and hence the theory of Markov chain is the mathematical procedure used for determining the probabilities of transitions from one state to another; and a *growing society model,* which consists of building up from a very small number of *n* members by adding members in succession. As each member is added to the society, he engages in contests with existing members to determine the dominance relation between them. Landau's conclusion for our interest is that social factors represented by the second and third models more easily yield "nearly hierarchical structure" than do those of the first model. Brams has utilized the concept of hierarchy in applications of several computer techniques to hierarchical decompositons of political systems that he defines in terms of different transaction flows between nations.

From the field of architecture and design, we have already mentioned the value that investigations of the geometrical regularities of natural structures hold for architects and the obvious hierarchic aspect of repeated form generated in shells or layers. The specific study referenced implicitly employs the concept of modular hierarchy in the following quotation: "Efficiency, the dominant criterion in technology, is closely linked to operations of periodic character . . . periodicity in production processes and in industralization of the building trade . . . calls for more periodicity in operations, involving series of repetitive elements and joints. Periodicity of forms in building has, therefore, an important technological-economic aspect over and above the aesthetic one." (Burt 1966).

Morrison, in one of the Kepes' *Vison Plus Value* series, further generalizes the concept of modularity as the basis of all order and diversity (Morrison 1966). His illustrations of modularity come from Chinese calligraphy, telemetered satellite images, modular aggregates of electronic circuits used in computers, cross-stitched embroidery, and programs for looms used in the weaving industry, as well as natural structures such as crystals, viruses, and giant biological molecules. His closing sentences suggest the magnitude of the importance of modular hierarchy, even though he does not acknowledge it. "The world is both richly strange and deeply simple. That is the truth spelled out in the graininess of reality; that is the consequence of modularity. Neither gods nor men mold clay freely; rather they form bricks." Ulam treats mathematically concepts of repetition by applying recursive relations to initial configurations of geometrical units such as squares or equilateral triangles. Patterns of growth derived from a morphological survey of elements combined according to these "recursive rules" in both time and space show an enormous variety of objects that are more complicated that the periodic patterns observed in crystals and other structures (Ulam 1962, 1966).

We also find examples of hierarchical structure in design methodologies (Alexander 1966, Manheim 1966). Alexander's *Notes on Synthesis* gives a detailed discussion of both the decomposition of a design problem and its recombination in solution. His point that "design" is more than "selection" (which can be treated by computer analysis) rests on the argument that for problems requiring "design" there exist no adequate descriptions of a range of alternative solutions nor criteria for evaluating these solutions in terms of the same descriptive symbolism. Again we encounter the problem of parts and wholes discussed under physics and biology. Alexander outlines a method for decomposing a problem into sets of "highly non-interacting" subsystems. The dilemma of decomposition is also found in the content of cybernetics research. Findeisen, in a discussion of optimal control in multi-level systems, summarizes the problem as follows: "The way to decompose a system is obvious if the subsystems can be formed so that they have no variables of the original system in common; this would mean that the original system is, in fact, composed of several non-interacting systems contributing to a common goal and may be subjected to a common resource constraint." (Findeisen 1968). Lasdon (1968) further discusses decomposition in mathematical programming. A succinct formulation of the general approach to the design of hierarchical systems from the point of view of optimal control theory is given by Pearson: "Large organizations of economic or biological nature inevitably appear to have a hierarchical chain of command. Characteristic of such organizations is that the hierarchy is a pyramid-like structure of decision problems and goals which vary in complexity. Problems at the base of the structure are usually fairly simple though numerous. Each of these is solved relative to a few *intervention* parameters which are themselves manipulated by higher more complex considerations. This structure of parameterized subproblems repeats itself up the hierarchy until at the apex there is one sophisticated problem upon which the outcome of the whole system depends" (Pearson 1966).

We next move to the subject of cognitive processes and pattern recognition. Weyl's discussion, "Chemical Valence and the Hierarchy of Structures" (1949), might well have come under the category of hierarchy in nature, yet we place it here because of its relevance to "levels of knowing." Weyl's insight in this short discussion is the realization that understanding is a progressive series of descriptions at different levels. He illustrates this phenomenon with the graphic representation of chemical structure as developed by Kekule in 1859 through Sylvester's contribution in 1878 on to the deepest level of quantum-mechanical description developed in the twenties. His moral not to "take too literally such preliminary schemes as the valence diagram . . . yet even so have the courage to draw the lines firm" derives (according to Weyl) from Nicolaus Cusanus, who stressed that "if the transcendental is accessible to us only through the medium of image and symbols, let the symbols at least be as distinct and unambiguous as mathematics will permit." (Weyl 1949).

In a review essay on the current status of physics, Toulmin (1967) underlines many of the doubts and uncertainties that currently plague much of the fundamental theory. He characterizes two chief recurring difficulties that arise whenever a "limited repertory of units or atoms is invoked to explain a multiplicity of phenomena—the problem of interactions and the problem of levels." The currently accepted description of fundamental particles and the strengths of their relative interactions derives from experimental evidence that involves "bombarding material targets with progressively more penetrating beams." However, there are now "suspicions that the more transitory and uncommon of the 200-odd known fundamental particles may represent artificial by-products of our bombardment of matter." Many of the difficulties of late 19th century physical science were subsumed under the "change of level" that resulted from adopting the quantum-mechanics (discussed by Weyl). Toulmin asks if the same situation might not exist today and answers in the affirmative, "there are reasons for thinking that the changes in store for us may be quite drastic, . . . the year 1966 saw a revival of speculation about the form which these changes may take." (Toulmin 1967).

In the introduction to a collection of essays that discusses the epistemology and methods used in social science, Ando emphasizes the difficulty of identifying *causal* relations in social phenomena (Ando, Fisher and Simon 1963). The common theme of these essays concerns the question of using characteristics of *exact* hierarchical systems in understanding the *approximately* hierarchical structures found in social situations. He suggests that the "answer depends crucially on the time period over which the system is observed and on the closeness of approximation to the hierarchical structure . . . the closer the system is to the exact structure required, the longer the time interval over which the accuracy of prediction will be maintained." He is here referring to the economic statistical methods that have been developed within the past twenty-five years. Ando quetions the validity of these methods since they "presuppose an exactly hierarchical system and . . . systems generating economic data are not likely to be exactly hierarchical, but only approximately so." Platt also raises epistemological difficulties in studying hierarchically complex systems: " . . . the higher levels of organization must be consistent with the lower ones but are not necessarily predictable from them, any more than a "systems phenomenon" like a traffic jam – or the absence of one – is predictable from a complete knowledge of the physics and chemistry of an individual automobile and its driver." (Platt 1969).

From current different disciplinary directions, we find similar expression of concern with methods of knowing that result in age old dichotomies of subjective/objective, holistic/reductionist, organic/mechanistic, and so on (Whyte 1949, Maslow 1954, 1967, Gutman 1964, Polanyi 1958, 1966, Langer 1967). Common to all such trends, we find the necessity to employ notions of level and hence hierarchical structure.

Maslow discusses the failure of the reductionist approach to characterize human personality. He suggests directions in which to seek a holistic technique that would incorporate psychological data and be more adequate. He describes a hierarchic "clustering" technique that utilizes "levels of magnification" based on the fundamental concept of "being contained within" rather than of "being separated from." (Maslow 1954, 1966). In his "Theory of Metamotivation" which summarizes a life-long research effort into characterizing "self-actualizing" individuals (i.e., those who function at their full potential), Maslow employs hierarchical notions of ordered levels of needs and gratifications. He also suggests it is not necessary to call holistic effects super-natural because, "Not only is man part of nature, and it part of him, but also he must be at least minimumly isomorphic with nature (similar to it) in order to be viable in it. It has evolved him. His communion with what transcends him therefore need not be defined as non-natural or super-natural. It may be seen as a "biological" experience." (Maslow 1967).

Polanyi's "tacit knowing" achieves "comprehension by indwelling and all knowledge consists of or is rooted in such acts of comprehension." He discusses at length the hierarchy of levels found in living organisms and life's emergent quality in relation to the concept of tacit knowing. He argues that "it is impossible to represent the organizing principles of a higher level by the laws governing its isolated particulars." Since the hierarchies of levels found in humans such as conscious behavior and intellectual action are situated above that of the inanimate, it is necessary to posit a *principle of marginal control,* which is control exercised by a higher level on the particulars forming its lower level. This principle of marginality is "present alike in artifacts, like machines; in human performances, like speech; and in living functions at all levels." It removes the necessity for mechanical explanation of living functions that consist of explanation in terms of the laws of physics and chemistry (Polanyi 1966).

Gutman, in a monograph on the relationship of structure and function and its effect on the problem of behavior, claims that the reason for hierarchical arrangement of organic structure derives from assuming the primacy of the whole before the parts. He points out that "depending on whether one gives primacy to the parts or to the whole, one reaches different philosophical systems . . . the part-viewpoint leads to materialism, mechanism, and the admission of physical causality as the only legitimate explanatory principle . . . the whole-viewpoint leads inescapably to idealism in the widest sense, to an organismic approach, and to the inclusion of some directive or teleological principle." (Gutman 1964).

Langer in the first volume of her monumental work to construct a concept of "mind" also argues that to "understand life means to discover the differences between organic and inorganic matter . . . ." A subtle aspect of the question of parts and wholes is reflected in her recognition of the value of symbolic images, for "they, and they only, originally made us aware of the wholeness and overall form of entities, acts and facts in the world; . . . only an image can hold us to a conception of a total phenomenon, against which we can measure the adequacy of the scientific terms wherewith we describe it. We are actually suffering today from the lack of suitable images of the phenomena that are currently receiving our most ardent scientific attention, the objects of biology and psychology." (Langer 1967).

Whyte's program for a 'unitary science' speaks of a "language of process, supported by the authority of science, which can show man how to think if he is to understand nature and himself." (Whyte 1950). In similar vein to all of these citations, his elegant statement ". . . the penality for any principle which fails to express the whole is the necessity to co-exist with its opposite" is highly relevant to all ways of knowing.

Finally, under the concept of hierarchy in artifact, we come to classifications of knowledge and disciplines. Comte's classification of science in 1854 (Whyte in press) describes thought as a progression from "theological" formulations to "metaphysical speculation" to "positivistic" thought which is truly scientific. His hierarchical arrangement of disciplines results when any one discipline attains the level of positivistic thought; for example, mathematics was the first to attain this level; astronomy, physics and chemistry followed; biology is on the way; and eventually ethics and sociology will follow. In a more recent attempt of classify the subject matter of thought, Boulding (1956) outlines two possible approaches to achieving a general systems theory. The first is "to look over the empirical universe and to pick out certain general *phenomena* which are found in many different disciplines, and to seek to build up general theoretical models relevant to these phenomena. The second approach is to arrange the empirical fields in a hierarchy of complexity of organization of their basic individual or unit of behavior, and to try to develop a level of abstraction appropriate to each." The second approach includes the following levels of abstraction: 1) level of frameworks, 2) level of clockwords, 3) level of the thermostat, 4) level of the self-maintaining structure, 5) genetic-societal level typified by the plant, 6) animal level, 7) human level, 8) level of social organization, and 9) transcendental systems.

In a paper that has been reprinted several times, Gerard (1957) defines basic units and concepts in biology and emphasizes biosocial comparisons. In addition, he is also concerned with defining boundaries of disciplines and their objects of study. Drawing upon the unit of *org* defined as "those material systems or entities which are individuals at a given level but are composed of subordinate units, lower level orgs, and which serve as units in superordinate individuals, higher level orgs" he delineates the objects of study as orgs at different hierarchical levels. He plots along the ordinate these different levels from molecule to populations of organisms, their properties of "becoming, being, behaving" along the abscissa. The resultant map becomes an outline of scientific effort delineating both disciplines and their respective content.

SUMMARY

This selection of references to the literature has emphasized entries of two basic types: one, those references that explicitly discuss hierarchical structure, and two, those where the results of investigation take the form of hierarchical arrangements. In annotating the many varieties of hierarchical form illustrated in these diverse disciplines and subjects, we have not attempted to generalize criteria for what properly is hierarchical structure in contrast to what is merely resemblance to hierarchical structure due to spurious, accidental or perceptual factors. At this early date in the investigation of hierarchical structure in nature and artifact, this selection is offered as an indication of candidate references to the literature. A future selection might well categorize spatial, temporal, scalar, functional, . . ., hierarchies among the two basic forms illustrated here – that is, branched and modular hierarchies. A more definitive list must await further refinement of the boundaries of hierarchical structure.

ACKNOWLEDGEMENTS

For their generous help in submitting titles and references to this bibliography, I am grateful to Mario Bunge, V. Gradecak, Marjorie Grene, Herbert Gutman, Chauncey Leake, Abraham Maslow, M. Mesarović, William Parkyn, John Platt, Cyril Smith, Robert Williams, Lancelot Law Whyte, and Albert Wilson. The final selection, however, is my own.

REFERENCES

Alexander, Christopher. 1965. "A City is Not a Tree" *Architectural Forum.* Reprinted in *Design,* No. 206, February 1966, pp. 46-55.

——. 1966. *Notes on the Synthesis of Form.* Cambridge: Harvard University Press, 216 pp.

Ando, Albert; Fisher, Franklin M.; and Simon, Herbert A. 1963. *Essays on the Structure of Social Science Models.* Cambridge: The M.I.T. Press.

Bateson, Gregory. 1968. "Redundancy and Coding." Ch 22 in *Animal Communication: Techniques of Study and Results of Research*, ed. T. A. Sebeok. Bloomington: Indiana University Press.

——. 1960. "Minimal Requirements for a Theory of Schizophrenia." *A.M.A. Archives of General Psychiatry,* 2:477-491.

Becker, Joseph, and Hayes, Robert M. 1963. *Information Storage and Retrieval: Tools, Elements, Theories.* New York: John Wiley & Sons.

Beckner, Morton. 1968. *The Biological Way of Thought.* Berkeley: University of California Press, 200 pp.

Bertalanffy, Ludwig von. 1952. "Levels of Organization." Chapter 2 in *Problems of Life.* New York: John Wiley & Sons.

——. 1955. "General System Theory." *Main Currents in Modern Thought,* 11:75-83. 75. Reprinted in 1956. *General Systems Yearbook* 1:1-10.

——. 1968. *General System Theory.* New York: Braziller, 289 pp.

Boulding, Kenneth. 1956. "General Systems Theory—The Skeleton of Science." *Management Science* 2:197-208. Reprinted in 1956. *General Systems Yearbook* 1:11-17.

Bradley, D. F. 1968. "Multilevel Systems and Biology – View of a Submolecular Biologist." In *Systems Theory and Biology,* ed. M. D. Mesarović, pp. 38-58. New York: Springer-Verlag.

Brams, Steven J. 1966. "Transaction Flows in the International System." *Am. Political Sci. Rev.* 40:880-898.

——. 1968. "Measuring the Concatenation of Power in Political Systems." *Am. Political Sci. Rev.* 42:461-475.

Brams, Steven J. 1969. "The Structure of Influence Relationships in the International System." In *International Politics and Foreign Policy: A Reader in Research and Theory,* Rev. ed., ed. James N. Rosenau. New York: Free Press.

Bronson, Gordon. 1965. "The Hierarchical Organization of the Central Nervous System: Implications for Learning Processes and Critical Periods in Early Development." *Behavior Science* 10:7-25.

Buck, Roger C., and Hull, David L. 1966. "The Logical Structure of the Linnaean Hierarchy." *Systematic Zoology* 15:97-111.

Bunge, Mario. 1959. "Do the Levels of Science Reflect the Levels of Being?" Chapter 5 in *Metascientific Queries.* Springfield, Ill: Charles C Thomas.

——. 1960a. "Levels: A Semantic Preliminary" *The Review of Metaphysics* 8:396-406.

——. 1960b. "On the Connections Among Levels." *Proceedings of the XII International Congress of Philosophy* Vol. VI, *Metaphysics and Philosophy of Nature*, pp. 63-70. Florence: Sansoni.

——. 1963. "Levels." Chapter 3 of *The Myth of Simplicity,* Problems of Scientific Philosophy, pp. 36-48. New York: Prentice-Hall.

——. 1967. 'Partition, Ordering and Systematics." in Chapter 2 of *Scientific Research I, The Search for System,* pp. 74-96. New York: Springer-Verlag.

Burt, Michael. 1966. "Spatial Arrangement and Polyhedra with Curved Surfaces and Their Architectural Applications." Masters Thesis, Israel Institute of Technology. Haifa, 139 pp.

Caspar, D. L. D., and Klug, A. 1962. "Physical Principles in the Construction of Regular Viruses." In *Symposia on Quantitative Biology*, Cold Spring Harbor, Vol. XXVII, pp. 1-24.

——. 1963. "Structure and Assembly of Regular Virus Particles." in *Viruses, Nucleic Acids, and Cancer.* Seventeenth Annual Symposium on Fundamental Cancer Research, pp. 27-39. Baltimore: Williams and Wilkins.

Charlier, C. V. L. 1922. "How an Infinite World May be Built Up." *Arkiv för Matematik, Astronomi och Fysik*, Band 16, No. 22, pp. 1-34.

Chomsky, Noam. 1957. *Syntactic Structures.* The Hague: Morton and Company.

——. 1967. "The Formal Nature of Language." Appendix A in *Biological Foundations of Language* by Eric H. Lenneberg, New York: John Wiley & Sons, 489 pp.

Coxeter, H. S. M. 1963. *Regular Polytopes.* 2nd ed. New York: Macmillan. 321 pp.

Doxiadis, C. A. 1968. "Ecumenopolis: Tomorrow's City." *Britannica Book of the Year 1968*, pp. 16-38. Chicago: Benton.

Findeisen, Wladyslaw. 1968. "Parametric Optimization by Primal Method in Multilevel Systems." *IEEE Trans. on Systems Science and Cybernetics* SSC-4:155-164.

Forney, G. David. 1966. *Concatenated Codes.* Cambridge: The M.I.T. Press, 147 pp.

Fuller, Buckminster R. 1965. "Conceptuality of Fundamental Structures." From *Structure in Art and in Science*, ed. Gyorgy Kepes, pp. 66-88. New York: Braziller.

Gardner, Martin. 1966. "The Hierarchy of Infinities and the Problems It Spawns." *Scientific American*, 214:112-118 (March).

Gerard, R. W. 1957. "Units and Concepts of Biology." *Science* 125:429-33.

——. 1958. "Concepts of Biology." Edited Proceedings of the Biology Council, Nat. Acad. Sci.. Reprinted in *Behavioral Science* 3:92-215.

Grene, Marjorie. 1967. "Biology and the Problem of Levels of Reality." *New Scholasticism* 41:427-449.

Gutman, Herbert. 1964. "Structure and Function." *Genetic Psychology Monographs* 70:3-56.

Harrison, E. R. 1965. "Olber's Paradox and the Background Radiation Density in an Isotropic Homogeneous Universe." *Monthly Notices Royal Astronomical Soc.* 131:1-12.

Hawkins, David. 1964. *The Language of Nature,* San Francisco: W. H. Freeman and Co., 372 pp.

Hawkins, David. 1967. "On Understanding the Understanding of Children." *American Journal Diseases of Children* 114:513-20.

Huffman, David A. 1952. "A Method for the Construction of Minimum-Redundancy Codes." *Proceedings of the I.R.E.*, pp. 1098-1101.

Kellenberger, Edouard. 1966. "The Genetic Control of the Shape of a Virus." *Scientific American* 215:32-39.

Khailov, R. M. 1964. "The Problem of Systematic Organization in Theoretical Biology." *General Systems Yearbook* 9:151-57.

Koestler, Arthur. 1964. *The Act of Creation.* New York: Macmillan Company, 751 pp.

——. 1967. *The Ghost in the Machine.* New York: Macmillan Company, 384 pp.

Landau, H. G. 1965. "Development of Structure in a Society with a Dominance Relation when New Members are Added Successively." *Bulletin of Mathematical Biology* 27:151-60.

Langer, Susanne K. 1967. *Mind: An Essay on Human Feeling.* Baltimore: John Hopkins Press, 487 pp.

Lasdon, Leon S. 1968. "Duality and Decomposition in Mathematical Programming." *IEEE Trans. on Systems Sci. and Cybernetics* SSC-4:86-100.

Leake, Chauncey D. 1967. "Technical Triumphs and Moral Muddles." *Annals of Internal Medicine* 67:43-50.

Lucky, Robert W. 1967. "Information Theory and Modern Digital Communication." In *New Methods of Thought and Procedure*, eds. Zwicky and Wilson, pp. 163-199. New York: Springer-Verlag.

Manheim, Marvin L. 1966. *Hierarchical Structure: A Model of Planning and Design Processes.* Cambridge: The M.I.T. Press, 227 pp.

Maruyama, M. 1965. "Metaorganization of Information." *Cybernetica*, No. 4. Reprinted in 1966. *General Systems Yearbook* 11:55-60.

Maslow, Abraham H. 1954. "Holistic-Dynamic Theory in the Study of Personality." Ch 3 in *Motivation and Personality*, pp. 22-62. New York: Harper and Row.

Maslow, Abraham H. 1966. *The Psychology of Science.* New York: Harper and Row.

——. 1967. "A Theory of Metamotivation: The Biological Rooting of the Value-Life." *J. Humanistic Psychology.* 7:93-127.

Mesarović, Mihajlo D. 1968. "Systems Theory and Biology–View of a Theoretician." In *Systems Theory and Biology*, ed. M. D. Mesarović, pp. 59-87. New York: Springer-Verlag.

Miller, James G. 1964a. "Adjusting to Overloads of Information." *Disorders of Communication* Vol 42, Research Publications, Assoc. for Research in Nervous and Mental Diseases.

——. 1964b. "Psychological Aspects of Communication Overloads." In *International Psychiatry Clinics: Communication in Clinical Practice*, pp. 201-224, eds. Waggener and Casek. Boston: Little, Brown and Co.

——. 1965. "Living Systems: Basic Concepts, Structure and Process, Cross-Level Hypotheses." *Behavioral Science*, 10:193-237; 337-79; 380-441.

Morrison, Philip. 1966. "The Modularity of Knowing." In *Module, Proportion, Symmetry, Rhythm,* ed. G. Kepes, pp. 1-19. New York: Braziller.

Needham, Joseph. 1943. *Time, The Refreshing River.* London: Allen and Unwin.

——. 1945. "A Note on Dr. Novikoff's Article." *Science* 101: 582.

Novikoff, Alex B. 1945. "The Concept of Integrative Levels and Biology." *Science* 101:209-15.

Palade, George E. 1963. "The Organization of Living Matter." In *The Scientific Endeavor*, pp. 179-204. New York: Rockefeller Institute Press.

Pearson, J. D. 1966. "Decomposition, Coordination, and Multilevel Systems." *IEEE Trans. on Systems Sci. and Cybernetics* SSC-2:36-40.

Platt, J. R. 1969. "Commentary on the Limits of Reductionism–Part I." *J. History Biology* 2:140-147.

Polanyi, Michael. 1958. *Personal Knowledge*, Towards a Post-Critical Philosophy. Chicago: University of Chicago Press, 428 pp.

Polanyi, Michael. 1966. *The Tacit Dimension*, New York: Doubleday and Company, 108 pp.

Prosser, C. Ladd. 1965. "Levels of Biological Organizations and Their Physiological Significance." In *Ideas in Modern Biology,* ed. S. A. Moore, pp. 359-360. New York: Doubleday.

Purcell, Edward. 1963. "Parts and Wholes in Physics." In *Parts and Wholes*, ed. Daniel S. Lerner, pp. 11-39. New York: Free Press.

Rapoport, Anatol. 1966. "Mathematical Aspects of General Systems Analysis." *General Systems Yearbook* 11:3-11.

Redfield, Robert. 1942. Introduction in *Levels of Integration in Biological and Social Systems*, pp. 5-26. Jacques Catell Press, Reprinted in 1968. *Modern Systems Research for the Behavioral Scientist*, ed. Walter Buckley, pp. 59-68. Chicago: Aldine.

Roosen-Runge, Peter. 1966. "Toward a Theory of Parts and Wholes: An Algebraic Approach." *General Systems Yearbook* 11:13-18.

Rosen, Robert. 1967. *Optimality Principles in Biology*. New York: Plenum Press, 198 pp.

——. 1968. "Some Comments on the Physico-Chemical Description of Biological Activity." *J. Theoret. Biol.* 18:380-386.

Sankaranarayanan, A. 1969. "On a Group Theoretical Connection Among the Physical Hierarchies." Research Communication No. 96. Douglas Advanced Research Laboratories, Huntington Beach, California.

Shapley, Harlow. 1958. *Of Stars and Men.* Boston: Beacon Press, 157 pp.

Simon, Herbert. 1962. "The Architecture of Complexity." *Proc. American Philosophical Soc.* 106:467-82. Reprinted in *General Systems Yearbook* 1965, 10:63-76, and *The Sciences of the Artificial.* 1969. Cambridge: The M.I.T. Press, 123 pp.

Smith, Cyril Stanley. 1954. "The Shape of Things." *Scientific American* 190:58-64.

——. 1964. "Structure, Substructure, Superstructure." *Reviews of Modern Physics* 36:524-32. Reprinted in 1965. *Structure in Art and in Science*, ed. G. Kepes. pp. 29-41. New York: Braziller.

Smith, Cyril Stanley. 1968. "Matter versus Materials: A Historical View." *Science* 162:637-644.

Stewart, Robert M. 1963. "Fields and Waves in Excitable Cellular Structures." In *Self-Organizing Systems 1963*, ed. James Garvey, Office of Naval Research, U.S. Government Printing Office, ACA-96, 78 pp.

Toulmin, Stephen. 1967. "The Physical Sciences." In *The Great Ideas Today 1967*, eds. R. M. Hutchins and M. J. Adler, pp. 159-195. Chicago: Benton.

Ulam, Stanislaw. 1962. "On Some Mathematical Problems Connected with Patterns of Growth of Figures." In *Proceedings of Symposia in Applied Mathematics*, Am. Math. Soc., Providence, Rhode Island, 14:215-24.

——. 1966. "Patterns of Growth of Figures: Mathematical Aspects." From *Module, Proportion, Symmetry, Rhythm*, ed. G. Kepes, pp. 64-73. New York: Braziller.

Von Neumann, John. 1958. *The Computer and the Brain*. New Haven: Yale University Press.

——. 1966. *Theory of Self-Reproducing Automata*. Urbana: University of Illinois Press, 388 pp.

Weaver, Warren. 1948. "Science and Complexity." *American Scientist* 36:537-44.

Weiss, Paul. 1967. "One Plus One Does Not Equal Two." *The Neurosciences*, eds. Querton, Melnechuk, Schmitt, pp. 801-821. New York: Rockefeller University Press.

Weizsäcker, C. F. von. 1951. "The Evolution of Galaxies and Stars." *Astrophysical Journal* 144:165-86.

Weyl, Hermann. 1949. "Chemical Valence and the Hierarchy of Structures." Appendix D in *Philosophy of Mathematics and Natural Science*. New Jersey: Princeton University Press. Reprinted in *Applied Combinatorial Mathematics*, ed. E. F. Beckenbach. New York: Wiley & Sons, 608 pp.

Whyte, Lancelot Law. 1949. *Unitary Principle in Physics and Biology*. New York: Holt.

Whyte, Lancelot Law. 1950. *The Next Development in Man.* New York: Mentor Books, 255 pp.

——. 1965. *Internal Factors in Evolution.* New York: Braziller.

——. 1965. "Atomism Structure and Form." In *Structure in Art and in Science*, ed. G. Kepes. New York: Braziller, 189 pp.

——. (In press.) "The Structural Hierarchy in Organisms." From *Unity and Diversity in Systems,* eds. Jones and Brandl. New York: Braziller.

Wilson, Albert G. 1965. "Olbers' Paradox and Cosmology." The RAND Corporation Paper P-3256, October, 20 pp.

——. 1967. "Morphology and Modularity." In *New Methods of Thought and Procedure*, eds. Zwicky and Wilson, pp. 298-313. New York: Springer-Verlag.

Woodger, J. H. 1937. *The Axiomatic Method In Biology.* Cambridge: Cambridge University Press.

——. 1952. *Biology and Language.* Cambridge: Cambridge University Press.

■

Epilogue

A characteristic of the current renaissance in epistemology is the intellectual thrust toward a more comprehensive and interrelated picture of nature, man, and society. In adopting a broad multi-disciplinary approach to the theme of *hierarchy*, this symposium explored what was felt to be one promising path toward such a coordinated view. In retrospect, the basic question relevant to this goal is whether the apparent structural analogies, all labeled with the term hierarchy, do indeed converge toward a single representation. The point of departure of the symposium was the focus on structure and function as essence, with atoms, cells, stars, and codes taken only as alternative mediums for the expression of the essence. While the basic question is what, if any, properties of hierarchies are medium independent, an important corollary question is what analogous structural and behavioral patterns display confluences sufficient to allow the formulation of precise propositions valid over the set of specific hierarchies entering the confluence.

In answer to these basic questions, we may cite such propositions as: *A stable aggregate will form only if its elements interact in such a way as to modify their internal structure* and *Hierarchical organization requires a balance between the number of degrees of freedom of its elements, the number of fixed constraints which function as a record, and the number of flexible constraints which program its evolution.* These propositions are nearly medium independent and indicate that there do exist hierarchical concepts of broad applicability. The extent to which they may be precisely formulated remains to be seen. Less broadly, the symposium exhibited evidence that analogies between hierarchical phenomena within certain clusters of disciplines, especially the bio-social-computer cluster, took on greater richness indicating that more intensive and detailed study within such a confluence should prove fruitful.

A second predication of the symposium's multi-disciplinary approach to hierarchy was the usefulness of analogy, however tenuous. While analogies range from those rich and deep enough to become the basis for productive and predictive theories, to those too superficial to provide even specious illustrations; whatever their validity, analogies constitute a basic mode of epistemological exploration. Through the simultaneous consideration of two or more analogous specifics, we are enabled both to parameterize and generalize. Hence, in the initial stages of investigating any specific hierarchy, bold and broad use of the analogies between many hierarchies is productive.

We conclude that the broad multi-disciplinary approach to hierarchy should be continued in the future. A too rapid narrowing of the jointly considered subject area would remove opportunities to stimulate our intuitions concerning whatever principles of unification that may reside in the alternate realizations of common structural and functional organization. While improving the precision of formulation is always an important goal in science, it must not be confused with narrowing the domain of discourse. But ultimately the nature of the relation between the specificity of formulation and the extent of the domain of discourse is itself a problem of hierarchy.

■

Index

Author Index

Subject Index